Reclaiming the Revolution

Extraordinary Adventures in Politics
and Leadership at the Inflection
Point of Industry 4.0

Stephen Barber

UNIVERSITY OF
BUCKINGHAM
PRESS

UNIVERSITY OF BUCKINGHAM PRESS,
AN IMPRINT OF LEGEND TIMES GROUP LTD
51 Gower Street
London WC1E 6HJ
United Kingdom
www.unibuckinghampress.com

First published by University of Buckingham Press in 2023

ISBN: 978-1-91564-377-3

CONTENTS

For Patrick, Francis and Marianne

Reclaiming the Revolution

Introduction:
Dawn of the Inflection Point

'Familiarity with today is the best preparation for the future.'

I

A world in which each of us has a digital twin would offer very different possibilities from today. It would offer extraordinary possibilities. And yet, we are fast approaching a time when everything that exists around us physically can be replicated digitally: you, me, each of us, our homes, societies, towns and cities, workplaces, even the planet itself. Existing in the metaverse, and powered by Artificial Intelligence and deep learning, these replicas can and will experiment, model, and help improve our physical world.

The potential of this technology is transformative. With our organs all replicated digitally, it will tackle disease, making us healthier and longer lived. Our towns and transport infrastructure will be planned to better suit our needs. Organisations will be more efficient, effective and performance strengthened. Products will be designed better and more cost-effectively. Communication will take place on a more immersive level. With a virtual twin who is less easily tired or distracted, our own productivity and creativity will be magnified. It is a step towards what Klaus Schwab, founder of the World Economic Forum, described as 'a fusion of our physical, our digital, and our biological identities'.[1]

DestinE (or Destination Earth) is the European Commission's ambitious large-scale development to create a digital Earth. In collaboration with the European Space Agency, DestinE will model and simulate our planet's systems. It will mitigate the impact of natural disasters, manage water, improve food production, and understand the impact of human activity. It will model, forecast, visualise, test scenarios and ultimately help solve the biggest challenge facing the world: climate change.[2]

This technology is amazing, of course it is. Not simply owing to the possibilities but because, alongside advances in AI, bio science, quantum computing, the internet of things, the internet of bodies, it represents change so profound, so significant, that it alters the trajectory of our societies, our communities, our workplaces, our economy and our politics. This is the Fourth Industrial Revolution. It is a moment in history. It is a transformative revolution that needs to be understood and acknowledged as the principal political challenge of our era. It is a challenge that requires transformed leadership and new ways of doing politics. And while digitally driven, it is a challenge that must be reclaimed for humanity.

Reclaiming the Revolution is about this inflection point.

II

In today's democracies there is a messy tussle playing out. It is a tussle between the forces of mainstream politics - who have existed in a sort of political market place for a quarter of a century and which no longer speak to a significant portion of the electorate – and the destructive forces of populism

– which threaten the stability of the system, the economy and society. While progressives cheer at each victory, big or small, the forces of populism prevail, dividing and damaging. Such unexpected polarisation has not only emerged at this inflection point, a specific reality that appears to concern so few commentators of our era, but more pointedly, is a consequence of the period of globalisation through which we have recently lived, a by-product of the third industrial revolution which brought us telecommunications and the microchip. The interconnectedness of globalisation has delivered great prosperity and opportunity. But it has also meant that so many ordinary people, voters, in post-industrial towns and cities of the developed world, feel left behind, abandoned almost by mainstream politics. And so, both sides of the political divide find themselves fighting the battles of the past and generating arguments of frustrating simplicity.

And the point is that neither side seems to have any convincing answers to the questions of the revolution before us. No, politics seems to have barely acknowledged the existence of the challenge. But the challenge is immense. What is this revolution for? Who will use it? Will it overwhelm humanity? Will it be democratic? Will it provide opportunities? Will it address the other challenges of our age? How can we prepare for it so that we can all benefit from its possibilities? How can we ensure this new world is inclusive?

You will see how this should be accepted as central to politics and policymaking, central to decisions we make about the economy, society, education, housing, transport, investment. And yet, it barely features in public discourse. Politics should be discussing and debating, it should be deciding how to harness the transformative power of this so that this, the

Fourth Industrial Revolution, is not something done to us, but something shaped by our values and ambitions.

But to do this, our politics not only needs to prepare for the dramatic transformations of the future, it also needs to really understand the present and that means appreciating the episodes, stories and developments of our past which have conspired to create today. That is the ambition of this book.

III

Ethel Merman was born at 359 4th Avenue in Astoria, Queens, in 1908. The daughter of an accountant and a teacher, she would become a performer of stage, screen and audio recordings through the 1930s right up to her death in 1984. In many ways, she was the natural choice to open the 1939 World's Fair. This huge exhibition was built upon a 1200 acre site, once a giant ash dump, just a few blocks away from her maternal grandmother's house where Ethel first entered the world.

The World's Fair was about the future. It was optimistic and ambitious. Physically, its great Trylon and Perisphere structures dominated the site, reaching into the air. But its genesis was in the experiences of the present. The 1930s had been a tough decade for America. The Great Depression had brought hardship and degradation for millions. Now, the Fair's organisers felt, was the time to be entertained and to look into a brighter future.

Admission, for those who wanted to attend the opening on April 30, cost 75c, and over the coming weeks some 44 million people passed through the gates, to gaze at the exhibitions. Merman's job was simple. She sang the Fair's song. Written by George Gershwin, it shared its title with the show, 'Dawn of a New Day':

Come to the Fair! (To the Fair, To the Fair)
It's the Dawn of a new day!
Sound the drats, roll the drums.
To the world of tomorrow we come!
From the Sun through the gray
It's the dawn of a new day!
Here we come young and old
Here to watch the wonders unfold
And the tune that we play
Is the Dawn of a new Day!
Tell the world (And the Door)
That we don't want to ground anymore
Better times here to stay
As we live and laugh the American way
Listen up one and all
There could be no resisting the fall
But tis the dawn
Of a new day!

Optimistic, cooperative, democratic, and egalitarian, there was a tacit recognition that the world was at an inflection point and it was one which would have to wait for a bloody world war before it could be realised. The new day was to be the space race, the atomic age, the superpowers, digital revolution of the 1960s and 70s, computers and the information age. And it all built upon the first industrial revolution which created the world we recognise today.

The World's Fair 'Dawn of a New Day' pamphlet declared a simple dictum: 'Familiarity with today is the best preparation for the future,' it told its excited readers. That could be the maxim of this book.

IV

We once again find ourselves gazing at the dawn of a new day, or more acutely, the inflection point of a new great adventure. It is one which will mark a change in our world as profound as that from agrarian to industrial society in the seventeenth century and from the Great Depression of the 1930s to the golden age of capitalism, eyed so optimistically from the World's Fair. But it is also a great adventure grounded in the disruption of today.

An inflection point is some significant transformation, event or development that changes or hastens the trajectory of an organisation, an industry, a society, a country, an economy or a geopolitical system. It can be seen as a good or a destructive phenomenon sparked intentionally or by accident, but it is a change so significant that circumstances are altered permanently. It changes our environment, our behaviour and the way we think. The inflection point we find ourselves reaching today is so significant that it has the power to transform our organisations, industries, workplaces, societies, nations, economics, political discourse, leadership and geopolitics. Think about the potential power of DestinE and our digital twins and consider that this is just one tiny development in a sea of transformative digital invention. Consider that Artificial Intelligence built to discover new lifesaving drugs can be just as proficient at creating new deadly warfare agents.[3] Welcome to the inflection point – it is time we started talking about how we want to harness its great power.

In a sense that is what visitors to the World's Fair, listening to Ethel Merman's stirring rendition of 'Dawn of a New Day', were being invited to do. Because the Fair did not simply showcase invention from across the world, it went much further. In this period of economic hardship, the World's Fair also promoted the ideas and values which might shape the new era.

And as we consider our own inflection point, the path which led us here and the state of our world today, we might reflect on the values, ideas and humanity that we wish to preserve and to promote. We might consider the great adventures that led us to this point and the lessons learned for our society, economy, politics and the sort of leadership we need. With this in mind, there are three factors which could be said to represent the most critical intersection of our disruptive age and which hold out warnings and opportunities for the age to come. These are the extraordinary adventures in the Fourth Industrial Revolution; extraordinary adventures in politics; and extraordinary adventures in leadership. It is around these big themes that this book is based.

V

Why write a book about this? Because too much of the current exploration is stuck in a sci-fi world of technology when the role of humanity is so much more vital and so overlooked. And because so much of contemporary debate fails to move beyond the parameters of today without learning the lessons of yesterday. This book is set firmly in the extraordinary adventures of our time, looking into the future and reaching into the past to understand industrialisation itself. There are stories big and small, connected and disparate. But each, in their own way, are about humanity and values as much as technology. Some are positive and uplifting, others dark and concerning. Perhaps what they all highlight, however, is that at this inflection point we need to understand our values more than any time before. The reason Ethel Merman was asked to perform back in 1939, after all, was because of humanity not technology. It was about familiarity not uncertainty. We must not forget that lesson today.

The Fourth Industrial Revolution really will revolutionise our way of life, and while it will alter our politics, it is essential to recognise that our politics is the only vehicle available to shape the revolution, to harness its power and to imbue it with the ideas we value. This is a deeply political book that raises questions about our capitalist systems, our democratic processes, our leadership and why decisions are made in the way that they are. Humanity must not be some sort of passive observer to this inflection point; the digital revolution must be shaped by us.

We need to accept the determinism of a book like this. It is possible that we create some sort of self-fulfilling fantasy, and that is a danger in all the enthusiastic analysis of what is about to befall us. That is, the mountain of opinion that foresees a particular future – fantastical, digital and transformative – makes that future more likely. The more that influential opinion anticipates what this inflection point will look like, the more that businesses, public bodies and people will prepare for it. The more we do this, the more likely it is to happen. And it is for that reason that the book is organised around these three big themes and that each of these themes is grounded in our own human experiences. It delves into the stories that shaped our world and connects them in often unforeseen ways. That is why this is a book that segues from Ethel Merman into the Coronavirus, and from eighteenth-century inventions to Donald Trump's rhetoric and then into advanced AI, with as much purpose as ease.

This is a book peppered with the real adventures of our times which together paint a picture of the future and allow us to understand the past. After all, the genesis of all that we know is in the present or the past. These adventures hop from country to country, from continent to continent, from century to century and decade to decade. This is not a dystopian treatise; it is optimistic, enthusiastic and realistic. These adventures are intended to provoke and to inspire.

This future has been foreseen for decades and there has been lots written about it. The difference today is that the future is upon us, it is a revolution, we need to understand it, prepare for it and respond to it. More than that, this is a revolution that must be reclaimed for humanity, for democracy and for all of us.

Reclaiming the Revolution is also a call to arms

PART 1:

EXTRAORDINARY ADVENTURES IN THE
FOURTH INDUSTRIAL REVOLUTION

What a Sexist Robot and Suicide Have to say about Industry 4.0

'It's like you guessing or flipping a coin'

I

Pepper has large black eyes with a periwinkle glow and a porcelain white face which is friendly and innocent. Pepper will engage in conversation with you, speaking in a cartoon-like way, gesticulating throughout with ten nimble fingers. Standing at just four feet tall, Pepper is unthreatening but hard to ignore. Pepper is, of course, a robot.

Pepper was born in 2014, created by the French firm Aldrerban Robotics before becoming part of the Japanese multinational Softbank, which had been a partner through development. And it is in Japan that Pepper has become part of the here and now. The first 1,000 sold out in the first minute of going on sale, with new owners parting with 198,000 yen for each unit. Pepper's semi-humanoid form is intended to facilitate interactions with people and with the 10-inch display seemingly hanging round Pepper's neck, there is more than a passing resemblance to Twiki the cheeky 'Ambuquad' of *Buck Rogers*.

And like Twiki, this real android is designed not simply to respond to commands but to actually read human emotions. This is becoming a commonplace AI capability and unsurprisingly is being deployed to sales activities where the technology can monitor and assess customer mood, engagement or sentiment.[1] In a similar way to people, Pepper listens to tone of voice and scrutinises facial expressions. The idea is that this is not a simple programming trick but that Pepper will learn from the personality of its owner, meaning that its own 'personality' will develop over time. And this means that each of those 1,000 units sold in the first minute or the subsequent tens of thousands sold across the world will all behave a little differently. In time technology will learn, adapt, 'think' and act. Pepper is part of the digital transformation which is disrupting and reshaping our world.

II

The Fourth Industrial Revolution, or Industry 4.0, is what is happening today and what it means for the future. It is the automation, digitalisation and artificial intelligence represented by everything from Pepper the robot to the interconnectivity and internet of things, the possibility of biotech, big behavioural data and machine learning.

This chapter rethinks how technology is revolutionising how we live and work – in a similar way to the first industrial revolution. But rather than viewing this from the standpoint of the technology itself, it argues that today it is artificial intelligence that is shifting fundamentally what it is that humans can contribute to society and the economy. The inflection point we are reaching suggests wholesale change in the way we work, produce, interact and grow. It means that we will organise ourselves differently and our role in life is less certain. But humankind is not redundant and this is not a passive process. Indeed, those uniquely human skills are all the more crucial in this world. Humanity matters more. And those human skills needed in the coming decades include adaptability, leadership, creativity. It is here that new economic and social value will be created. AI is so often portrayed as a story about what technology can do. But it should also be a story about what people can do, about what technology can help humans to achieve. Technology cannot be something which is done to us against our will but rather a great opportunity to unleash human possibility.

III

Pepper is designed as an open platform, and this is one of the reasons why the robot has proved so attractive. Already Pepper can be found taking your order in restaurants, greeting you at office receptions and helping you to buy a new motor in car showrooms. You see, Pepper is much more than a 'plug and play' device and is better thought of as a canvas onto which innovation can take place. The vision is that 'the developer community will progressively sustain Pepper's growth.'[2]

Two people with an idea were Gurch Randhawa, Public Health Professor at the University of Bedfordshire, and his colleague Dr Chris Papadopoulos. Among many interests, Gurch and Chris had been concerned about staffing in care homes, the places that look after the elderly when they cannot take care of themselves.

Western populations are aging. Medical advances and lifestyles have extended lives while fertility rates have declined. That is, the elderly of today are living longer than their parents but had fewer children themselves. The result is that the fastest-growing demographic is the retired, with fewer younger people in work as a proportion to support them through their taxes. Meanwhile that aging population demands more and more medical and social care. Our demand for healthcare can be thought of as a U-shaped chart (with a long flat bottom). For most people, there is a need for resources at birth and early childhood, but that rapidly tapers off such that most of us need little attention until we get into our 60s or early 70s. But then the demand increases rapidly, and those in their 70s, 80s and 90s (and 100s) consume vast amounts of health care and social care. And it is care homes which are dealing with this on a day to day basis.

Short of resources with stretched staff, the problem in many homes is that with all the routine tasks of changing sheets, clearing up, tidying, bringing food and the like, staff simply do not have time to talk to those they care for. And yet that essentially human interaction is so desperately needed by many elderly; it is what caring is about. And it is all the more important when one considers that rates of depression and suicide are relatively prevalent amongst those residents in care homes. While still rare to completion, a systematic review conducted in 2014 showed just how common suicidal thoughts in care residents are and the suggestive evidence of the importance of the care environment including staffing has on this.[3]

So imagine if a low cost resource could be introduced to care homes which would mean those tasks get done but the elderly enjoy much more interaction and attention to their needs. For Gurch and Chris, Pepper was an obvious solution and they set about developing a robot to perform these functions. 'We put social robots in care settings to evaluate what additional impact they might produce alongside the routine provision of care, particularly in terms of loneliness and health-related quality of life for the older adults,' explains Chris. 'We felt this was vital given how stretched the care sector is, with care staff struggling to cope with demand and experiencing poor morale'.

But here is the rub. You won't find Pepper changing sheets or cleaning dishes. Pepper is much more capable than that. Pepper has been developed to talk to, interact with, entertain and amuse the elderly, while its human colleagues get on with the heavy lifting. Pepper has been developed to learn about what the old people being cared for like, talk to them about those interests, whether it is the cricket score or latest soap opera, play music or get in touch with their family. Gurch and Chris agree: 'In the future, social robots will become a staple part of enhancing care home residents' mental health though enjoyable, meaningful and sustained conversation while also alerting staff to problems arising or emergencies so that early human intervention can be achieved'. Pepper is today's revolutionary technology, but we have been here before.

IV

The world's first public telegraph company was founded in 1846 by the inventor William Fothergill Cooke and financier John Lewis Ricardo. Cooke had created an electric telegraph system capable of transmitting messages across distances of several miles using indicating needles moving electromagnetically. It was first deployed commercially in 1838 on the Great Western Railway between London's Paddington Station and West Drayton 13 miles away. In return for replacing inefficient underground cables with those suspended in the air on poles (at his own expense) and giving the Railway free use, the GWR freed Cooke from exclusivity. This allowed him to open public telegraph offices, and new railways adopted the technology as fast as they were built.

The power of this technology was illustrated to great attention on 1 January 1845 when the murderer John Tamwell escaped aboard the 7.42 from Slough, only to be apprehended on his arrival in London thanks to a telegraphed message. Hanged by the neck for his crimes a few weeks later, Tamwell will forever be remembered as the first person arrested because of telegraph technology.

Ricardo's finance later that year transformed Cooke's patented technology into the Electric Telegraph Company, established adjacent to the Bank of England in Founders' Court. While not exactly profitable, the company supplied railways, newspapers and stock exchanges, meaning that within a decade it was sending more than 700,000 messages each year and the cost of electronic communication was tumbling.

But none of this would have happened without industrialisation: the revolution which began in Britain in the eighteenth century and transformed the world from hand production to mechanisation.

Consider this. The political economy we know so well was forged in industrialisation, beginning less than 300 years ago. This is such a short period in human existence that there is no more reason to suppose it will continue indefinitely in its recognisable form than to speculate about the possible alternative futures. For most of human existence (let us say three million years since Palaeolithic age or when modern homo sapiens emerged 300,000 years ago or 50,000 years since behavioural modernity or even 10,000 years since we began sedentary agriculture), for most human beings, the world they died in had not changed fundamentally from the one into which they had been born. Consider that next time you visit a natural history museum and see hunting flints dating back a million years and those dating back 350,000 years. They are not much different. Hunter-gathering sustained humans for 90% of our existence on earth. Technology was slow to materialise even after more settled societies developed. The earliest discovered boat, the three-metre long Pesse canoe built around 8,000 BC, is not that much more primitive than those used by Native Americans 3,000 years later. You get the point.

For everything else that happened, whether it was the Roman Empire or merchant capitalism, it was the industrial revolution which changed all that.[4] This was the rapid growth of factories and production, capitalist economics and urbanisation. It was the extraordinary explosion in technology. That included communication such as the telegraph.

By the time that Cooke patented his telegraph technology, the second industrial revolution was underway. Beginning in the mid-nineteenth century, this saw electricity, communications and gas power drive a further economic and social expansion. It was an extraordinary transformation in the way we live.

One area which changed so rapidly as a result of industrialisation is the world of work. And yet we sometimes consider this as something rather stable. But the very phenomenon that facilitated the first industrial revolution and the second (and third which saw the growth of the microcomputer, electronics and mass communication) continues to change how work is done. Now we are at the inflection point of it revolutionising our world once again.

For perhaps even more than 500 years before industrialisation, there was a feudal system in places like Britain. This was a medieval economic model based on peasantry and the manor. For the majority who did not have the good fortune to be barons or knights, life was a tough slog, farming for their existence on a small amount of allotted land. Human existence was rural and basic.

Technology changed that (in part), making farming less labour-intensive, producing more food with fewer people and freeing up the productive capacity of those people to do something else. People moved away from the farms and manufactured new products in the factories that sprung up in the rapidly growing cities of the country.

Since then, the shift has consistently been one from labour intensity to technological intensity. And some jobs which were created through industrialisation have themselves become obsolete. As light came to our streets in the eighteenth century, it was a crucial job that someone would go around lighting and maintaining these candles or gas lamps. Lamplighters were of course replaced when automatic gas, and then electric lamps started to be used. In the nineteenth century many offices had a 'computer', often a young woman, employed to do calculations. This job became unnecessary mainly because of, well, the computer. Think about the difference between a modern car manufacturing plant and the production line established by Henry Ford. Today more cars can be produced, to a much higher standard and with far fewer workers. And yet all of this has not meant mass unemployment. That phenomenon which freed up productive capacity at the start of industrialisation has done the same again and again as unskilled and some semi-skilled work is replaced by automation.

Industrialisation really was a revolution. It was not all about economic and technological advancement but meant social and political change too. Indeed the social and political change was just as much a part of the revolution. And the technology that was being created could be used for more than commercial gain.

Government used its reserved powers to control the Electric Telegraph Company in 1848 as the radical Chartists demanding democratic reform for working men found their communications disrupted. And in 1868 the Telegraph Act nationalised the ETC and other telegraph companies, taking them into the ownership of a state monopoly under the General Post Office, and there it stayed until becoming a separate department known as Post Office Telecommunications in 1969. By now we were experiencing the third industrial revolution. A transformation in digital, computing and communication technology and we were in a world recognisable to most of us today.

But something different has started to happen as we approach Industry 4.0. Technology is getting so good that it has the potential to perform skilled work and maybe (much) better than human beings. That means driverless vehicles, business administration, accountancy, actuarial work, investment analysis, legal analysis and even journalism. Indeed, in February 2020, Reuters used an AI prototype to deliver an automated, presenter-led sports report. It offered the potential for fast, efficient reporting and personalisation of news.

This all not only means that fewer people will be needed to perform these functions in the future, but also that the great swathes of middle management, the comfortable staple of most organisations today, could well be near obsolete.

Cooke and Ricardo could hardly have imagined what their venture would evolve into. Post Office Telecommunications became British Telecom by Act of Parliament in 1981 and the 1868 Act was repealed in 1984 when the company was privatised by the government of Margaret Thatcher. At the time a flotation of even half of BT was the biggest listing the London Stock Exchange had seen. Advances in telecommunications forced change with a loss of monopoly, new competition and new technologies transforming the landscape. In the three decades since privatisation, BT has bridged the third into the Fourth Industrial Revolution in the technology it deploys and the way it goes about its business. Indeed, it signalled the way in 2018 when it announced a round of some 6000 redundancies and the creation of 4000 new posts. The jobs being lost were among the company's middle management. The new jobs were a polarised list of highly specialised technical positions on the one hand and on the other... call-centre work. The economy is being hollowed out.

V

But there are still jobs out there – more, in fact, than ever before. And if you have applied for a post recently, you might have noticed that the process is different from a few years back. In the past, an organisation wishing to fill a position might have posted the vacancy in the trade press, inviting suitable applicants to obtain an application form or submit their resume, which would be returned with a neatly typed cover letter by post. Having sifted through the responses, looking for evidence of those with the strengths to do the job, a manager might have made two piles of paper applications on her desk – one for shortlisted candidates and another for rejections. The lucky candidates in the first pile would have received a letter inviting them for an interview. The rest would have opened a polite 'thanks but no thanks'. It was based on a judgement of the person doing the sifting, and that judgement was usually pretty unscientific.[5]

Recruitment was one of the early business functions to be adopted by the internet. Websites were created making it easy for candidates to search for jobs, breaking them down by sector, skills required and the all-important pay offered. On the other side, candidates could match their skills, build CVs and apply for a hundred posts in the time it might have taken them to prepare a single application in the past. Today pretty much every sizable employer has an online application system. It makes the whole process more streamlined and transparent.

But technology has moved beyond process. AI is able to do all that reading of resumes, sifting, shortlisting and rejecting – all without human involvement. You would think that this is fairer. After all, the old system suffered from human prejudice: the manager didn't like your school, family background or age for instance, and you were rejected even though you might have done the job brilliantly.

Back in those pre-digital days, Marianne Bertrand, an economist at the University of Chicago, conducted a controversial field experiment. Trawling through the 'Help Wanted' adverts of newspapers in Boston and Chicago itself, she and her fellow researchers applied for the jobs using a series of fictitious CVs. But they used distinctly African American-sounding names on some resumes and distinctly 'white'-sounding names on others. Is there any reason to think that Emily Walsh would be a better employee than Jamal Jones?

Well, the shocking result was that white-sounding names were invited to 50% more interviews than black-sounding names. Applicants showing addresses located in wealthier neighbourhoods also enjoyed more callbacks. And the results were consistent across industries. The conclusion was that there was significant bias against assumed black candidates by the people sifting through applications.[6]

Industry 4.0 will involve recruitment that is much more sophisticated, objective and scientific. And this is important given the hollowing out of the economy. Organisations will need to be able to recruit capable specialists able to adapt to disruption and push the boundaries of what is possible.

The e-commerce giant Amazon would seem well placed to be a trailblazer in this area. Its strategy of automating sales and warehousing has evolved into a sophisticated business model which has left traditional competitors behind. It was a natural extension then that Amazon's machine-learning and AI specialists should turn their attention to recruitment, devising technology that would scour thousands of resumes and identify the top talent to hire. What they built was astonishing. It went further than simple algorithms. Amazon's creation would learn from its own practice about selection, performance and the 'best' talent. And this was no small endeavour. Between 2015 and 2018, the number of people working for Amazon globally more than tripled to almost 600,000 employees – that's comparable to the population of a city like Milwaukee.

The project was set up secretly and was put into operation evaluating applications to identify those high performers. And a point should be underlined. This was much more sophisticated than a developer identifying positive (or negative) attributes and the computer recognising these. Machine-learning algorithms meant that the technology would be able to teach itself to select the best candidates because it would be able to learn.

And it learned by examining huge historic data from resumes submitted over a ten-year period. It was able to use a star system to rank applicants in a way familiar to those who buy products from Amazon's website and review just how good it was. But the recruitment engine taught itself something that would ultimately be unpalatable to bosses at Amazon and lead to the programme being closed down. It learned (wrongly) that men were more successful than women. It discovered that most applicants were men and that most job offers were made to men. It became biased against women. The experience was no more objective than when Marianne Bertrand responded to Help Wanted ads in Chicago and Boston a decade earlier using African American names. Only this technology was even better at discriminating.

So the recruitment engine actively looked for clues as to the applicant's gender. This could have been attendance at an all-girl school or captain of the women's hockey team. It even looked for verbs more commonly found in men and women's CVs to differentiate. These factors made it less likely that a female applicant would be shortlisted.

Yochanan Bigman of the Hebrew University led some research which is insightful, and worrying, here. Coining the phrase 'algorithmic outrage deficit', he discovered that people are simply less morally outraged at discrimination that happens because of an algorithm than if a human being has been involved. People are less motivated to do anything about the discrimination.[7] And yet, as the situation at Amazon might illustrate, algorithms can emerge from extant or historic inequalities.

This is unfortunate because, separately, Amazon appears to really care about the gender balance in its industry and has gone to some lengths to support women. An example of this is its 'Women in Innovation Bursary' paid to women from less advantaged backgrounds to support ambitions to work in the field of technology. It included financial support as well as mentoring and help building CVs. The intention was bold: 'The bursary aims to help break down barriers to women entering the ... Science, Technology, Engineering & Mathematics workforce with the objective of improving and diversifying the gender balance in the workplace'.

The sad irony here is that any woman applying for a job at Amazon who proudly included winning an Amazon 'Women in Innovation Bursary' on their resume would have been downgraded by Amazon's gender-biased recruitment engine for doing just that. They would be less likely to get the job. The issue is with how the algorithm was built and its reliance on limited or biased data. 'Garbage in, garbage out' is the difficulty here, but it is maybe not enough to expect even improved inputs to be unproblematic.

VI

There is a lot that we know about suicide. That is the epidemiology of those who take their own lives, whether this be the young – where suicide is the second biggest killer of people aged 15-29 – or the elderly in care homes where Pepper has been developed to work. Large organisations such as BT invest heavily in health and wellbeing initiatives to prevent suicide in the workforce, sensing a responsibility to its staff.

The World Health Organisation considers suicide a serious public health problem, globally accounting for 800,000 deaths each year.[8] And for every successful suicide, there are as many as forty attempts.[9]

But there is still a lot more that is unknown, and that matters to doctors trying to do the best for their patients: their ability to intervene successfully. This is something which struck the psychologist Joseph Franklin of Florida State University. He set about examining half a century's worth of academic studies into suicide prevention.[10] And what he discovered was that we are simply not very good at predicting who is likely to attempt suicide. Worse than that, 'It's like you guessing, or flipping a coin is as good as the best suicide expert in the world who has all the information about a person's life,' he told *Medical Xpress*. 'That was pretty sobering for us and sobering for the field because it says all the stuff we've been doing for the past fifty years hasn't produced any real progress in terms of prediction.'[11]

What that meant was that, despite the 365 studies that he alone examined published year in year out since the 1960s, knowledge had really not advanced in the same way as other medical interventions. This was perhaps because the analysis was always limited to the identified risk factors such as mental illness, substance misuse or trauma. This, in spite of the vast number of studies and the huge amount of data available for scrutiny. It was limited to the ability of human researchers.

And so Franklin and his colleagues started working with artificial intelligence. They started to adopt similar expectations to those which had guided developers at Amazon when they built their recruitment engine, albeit with a dramatically different ambition. By applying machine learning to thousands of medical records as part of a huge database, they were able to produce algorithms that could accurately predict a suicide attempt close to 100%. That accuracy improved from 720 days to seven days before an attempt.[12] It was a massive leap forward.

Franklin's work has made a real difference. His algorithm is able to be implemented in hospital databases, meaning it can support potentially millions of patients. He has developed an app called the 'Tec-Tec' which it is hoped will reduce suicidal behaviours. And that should be the happy ending to this otherwise gloomy story. It should be that this AI tool has completely revolutionised suicide intervention and reduced massively instances of those taking their own lives.

But Franklin found that it has not yet been able to achieve its potential. And the reason for this is that the AI itself, of course, does not intervene. It merely presents the analysis to an attending mental health professional. And if what the machine is reporting does not accord with the professional instincts of a doctor standing before a patient, that doctor is likely to ignore the warnings.[13]

It is known as the 'black box problem'. The analysis going on inside advanced machine learning is so sophisticated that it produces results unobtainable by human inquiry. Unfortunately, that mystery makes expert humans distrustful of the outcomes rather than embracing of the possibilities this brings.

VII

Hanging proudly in Derby's Museum and Art Gallery for more than 130 years since it was purchased by public subscription is Joseph Wright's masterpiece *A Philosopher Lecturing*

on the Orrery. You will likely have seen it, since countless
engraved reproductions were made in the century after it
was painted in 1766. The original is sizable, one and a half
meters high by two meters wide and oil-painted in a realist
style which made Wright one of the great artists of the British
enlightenment.

The candlelit scene depicted in the painting was not without
controversy when it was unveiled since at the very centre of
the picture is neither a classical figure nor a religious subject
as was common at the time. No, at the centre of this painting
is the miracle of science, the Orrery; a model of the solar
system with the earth, moon and planets orbiting the sun,
the hidden light source in the painting. The Orrery is being
demonstrated by the authoritative lecturing philosopher.
Beside him is a student taking notes of which the philosopher's gaze appears to be checking. There are others looking
on in wonder – including the luminous faces of two small
children.[14]

Wright was depicting the spirit of the Industrial Revolution.
And this reminds us that it was not just technology that
made for this great transformation, or else it would likely
have started somewhere other than Britain. What made for
the Industrial Revolution was the Enlightenment; the coming
together of science and thought. Wright was a member of
Birmingham's Lunar Society, which met monthly to discuss
the latest in science, medicine, art and business (the light
flooding from the centre of the Orrery was surely a metaphor for this great possibility of becoming enlightened). The
Enlightenment included a political climate that embraced a
respect for reason and rationality rather than religious sup-
pression of earlier times or other countries. It was possibility

rather than tradition, liberalism over arbitrary authority. It included the concept of individual rights, articulated by John Locke, which permeated the old feudal order.

Let's not overstate this by modern standards, but at the dawn of the Industrial Revolution, Britain was notably the freest country in the world. Individual freedom of thought and action means the possibility of creativity and innovation. And this is the essence of capitalism: good and bad.

Industrialisation meant exploitation of the masses: workers in factories and mills producing while the owners profited. But the so-called 'great enrichment'[15] meant that these experiences eventually forced social change throughout society.

More than that, individual freedom represented a human value alongside action. Those values facilitated technological advancement, and that meant dramatic economic progress. Technology, then, released human capability.

In the painting, those viewers who looked on in awe at the scientific understanding did so in a climate of open thought and freedom. The two enigmatic children's faces looked on in the expectation that their world would be very different from that of their parents.

VIII

Kiichi Ishikawa, a sixty-year-old from the city of Yokosuka in Japan's Kanagawa Prefecture, made the headlines around the world in 2015 when he was arrested by police for unleashing a drunken attack on Pepper the robot. Security footage showed Ishikawa kicking the robot in a Softbank Corp store while the *Japan Times* reported the sad news that 'the damaged Pepper now moves slower'.[16]

Ishikawa's motives are unclear, but the obvious comparison is with the Luddites, the radical nineteenth-century English textile workers who destroyed the machine technology that replaced human craft. The inflection point at which humanity finds itself today promises changes as dramatic as those brought on by the first industrial revolution. And we need to be enlightened.

The economy is being hollowed out by this technological revolution. Jobs are being replaced, not just in unskilled work as has been the experience since industrialisation developed, but now robots, AI and machine learning are becoming capable of taking on skilled work. The natural suspicion is where this leaves us, people, humankind, and the fear that we will become redundant.

The full implications of this are open for debate (this book engages in them). But the predictions of the World Economic Forum should not be taken lightly. Its influential report on the future of work forecast that 7 million jobs are being lost globally to technology with just 3 million jobs being created because of it. [17] The idea that a three-year-old, like Marianne, just entering preschool is likely to enter a career that does not yet exist is a frightening prospect and one which naturally provokes the sort of response recognized by Ishikawa or the Luddites alike. But an optimistic analysis would conclude that humans are not being replaced by machines, only *some* human skills – the same as has long been true (albeit now higher skills and at an accelerated pace). This means that professionals entering the workforce might no longer need to be trained for middle-management roles, supervising and cajoling human resources to meet strategic aims of an organisation. Technology today has the power to free us from such burdens. Much more important today and tomorrow will be to foster these innately human capabilities of creativity, innovation, adaptability and leadership.

We also need to rediscover a truth that was known to members of the Lunar Society: a sense of our human values is essential. Of course the technology should be more efficient than humans – if that were not the case why on earth would we adopt it? It was as true in 1838 when the Great Western Railway adopted the telegraph as it is today. In the tentative discussions of today's economy, we so often ask the wrong questions. We ask how much more productive any given machine will be than the fallible human being it will replace. It is all ends not means. But means really matter. We should really be questioning the values (the human values) that drive that technology and how that cooperates with human skills and objectives. We should be asking what new human capabilities are unleashed.

Let us return for a moment to Amazon's recruitment engine and Joseph Franklin's suicide algorithms and ask why it was that they did not quite work as their developers had expected. Each case demonstrates convincingly just how technology can be so very powerful in scrutinising very big data. It is capable, beyond human analysis to find proxies for the markers that they are seeking (or in the case of women applicants to Amazon, avoid). Data which might seem irrelevant to a human analyst becomes pertinent when it correlates to other markers. The machine discriminates, even if that was not the intention of its creator; it learns, and that is why it can do so much more than programming of the past.

The weaknesses in these machine learning innovations were not technological. The weaknesses were human and they were failures of values.

Consider that Amazon's recruitment engine and Franklin's analysis of medical records were limited perhaps because they modelled only human behaviour not human values. In the case of Amazon the machine learned from how human recruiters had behaved in the past and replicated it at an accelerated pace. Essentially they learned incorrectly that behaviours of a kind that Marianne Bertrand uncovered were 'good'. When it came to the treatment of mental health patients and identifying those most at risk of taking their own lives, the limitation was the interface between the technology, which through the mystery of its 'black box' process was able to predict, and the behaviour of the human health professional who had to make a judgement about a patient.

The answer then is not less technology, for that is unstoppable. But it is not less human interaction either. Humans might be seen as the weak point of the chain in these developments, but what should be obvious is the need for more human collaboration. It is all about humanity and values. It is about the new possibilities of human capability.

The Industrial Revolution represented an enlightened knowledge transfer throughout society. At the inflection point of today, it should be clear that technology alone will be incapable of doing what was once the preserve of people. Artificial intelligence in isolation of human beings does not offer the greatest potential for technology. There must be more humanity and more collaboration between people and machines, especially in the workplace.

Pepper's future is not a lonely machine who works cheerfully but separately to the lower-skilled humans in the same organisation. Pepper's true potential is as a 'cobot'. AI technology that interacts with, works with and collaborates with human personnel. And human beings do not carry on as normal with Pepper assuming an ever increasing list of their (higher) duties. Humans can be transformed by this, capable of leading, shaping, innovating and creating. But only if we get our political, economic and social decisions right. And only with an enlightened sense of our values.

How Yesterday's Future Became Tomorrow's

'To preserve the benefits of what is called civilized life, and to remedy at the same time the evil which it has produced'

I

Imagine you were to receive a letter saying that, without obligation or taxation, you would be receiving €560 each month. It is not a life-changing sum of money, like winning the lottery or the Vegas Megabucks jackpot, but it would be guaranteed, a basic income. Just think about the difference it might make. How would it change your behaviour, your outlook, your life choices?

Liisa Koskinen received a letter which said just this. In fact she was one of 2,000 Finnish citizens selected at random to be part of a two-year experiment into the effects of a basic income.

Finland's experiment in a universal wage took place between 2017 and 2018. The idea was a regular unconditional payment made to citizens irrespective of whether they were in work or not. It was perhaps a way of tackling the country's relatively high unemployment problem. And perhaps surprisingly, the innovation gained support across the political spectrum. Here at least it seemed to offer the sort of state support favoured by the left but also the promise of welfare simplification and personal liberation to find new jobs or start new businesses that attracts the political right. A rare, and unusual, consensus.[18]

There is not much new about the idea of a universal basic income. The great thinker and activist Thomas Paine proposed it in the late eighteenth century. His pamphlet *Agrarian Justice* was written as an attempt to 'preserve the benefits of what is called civilized life, and to remedy at the same time the evil which it has produced'.[19] At that time large tracts of common land in Britain had been sold to a small number of wealthy landowners. This process was, of course, intimately connected to industrialisation. The growth of factories drew in workers from the countryside who moved in ever increasing numbers to the great cities, where there were new forms of work. Until this time not only farming but also grazing of animals, fishing, foraging and growing crops took place on common land. That is land where commoners had the right of access. Actually, that word 'commoner', today used pejoratively, originally meant simply someone who had a right to use such land. It was shared and used in a kind of egalitarian way. You can imagine this kind of basic, rural life organised around the land to which families were tied generation upon generation.

The production capacity that came with larger-scale agriculture and the corresponding urban growth of industry put paid to this world. The Enclosure Acts fed the revolution over a century, from 1750 stripping away that independent life supported by cultivated nature. These Acts, numbering 4,000 by 1850, converted common land into private property, concentrated in a few hands. It replaced an agrarian country with a land measured as capital.

It is not difficult to understand the social divisions this caused and the desire among thinkers to address inequities. Indeed, that idea of 'equality' emerged from industrialisation itself. Among other measures, Paine proposed a tax that would provide for the needs of those without land and this included fixed payments of £15 to every man and woman over the age of twenty-one in the country. He justified this as 'a compensation in part, for the loss of his or her natural inheritance'.[20]

If for no other reason, that is interesting because it is a different justification from more modern proponents on the political left who see a basic income as a more efficient form of wealth redistribution and a challenge to capitalism. Those arguments have remained firmly at the periphery of debate, dismissed by mainstream and market-orientated policymakers. And the case against has been pretty convincing: a universal basic income is incredibly costly, requiring a massive hike in today's taxation; it distributes cash to the already wealthy as well as the poor, whereas welfare has been traditionally based on need. Not only that, they say, it removes incentives to create wealth.[21] That is, a market economy needs individuals to put their capital at risk. It is what John Lewis Ricardo did when he financed Cooke's Electric Telegraph venture and it illustrates that (some) inequality has always been necessary in a capitalist economy.

But it could just be that these early experiments in offering a basic income, in New Zealand, the Netherlands as well as Finland and now even Wales, illustrate how the inflection point of today means so many longstanding debates are being reappraised. And when we go back to relearn some of the lessons of the past, it is possible to view afresh some of the challenges of tomorrow.

II

What is this chapter about? It centres on the debates of the new industrial age and how they reflect, differ and evolve from the past. It considers the implications of technological revolution for our economy, world of work, skills and education. It delves into our recent history to understand how we ended up here and what it means for where we are headed. It is a chapter about how the arguments which have raged since industrialisation need to be reinterpreted for the new age – including equality, economic incentives and creation of demand.

It starts with a crumbling edifice in the American Midwest.

III

The Richman Brothers building on East 55th Street, Cleveland, Ohio, stands like some sort of decaying monument to an earlier age. With so many of its window panes broken in each of its five floors, it is as imposing today as when it was alive. Then it was the self-styled largest manufacturer of men's clothing in the world and employed thousands of workers on a staggering 60,000 square meters of space. Today its presence is still arresting, but its qualities are a ghostly reminder of what was once a vibrant and successful factory. Eerie and strikingly sad, it survives without purpose; a glorious past but seemingly no future. Now abandoned, empty and cavernous, it is difficult to imagine the noise and vitality that its great walls once contained or how the light shone through the huge glass windows onto workers producing hats, coats and other garments; on the 18-metre cutting table, then the biggest the world had ever seen. It is difficult to imagine the bustle of deliveries bringing supplies and materials or collections of products which must have been a daily occurrence. The boarded-up entrance and overgrown facade makes it difficult to picture the throngs of workers who would have spilled out of the great factory doors at the end of their shift.

Cleveland was once a thriving industrial city, producing steel, manufactured products and vast patents in the early part of the twentieth century and in the decades after the Second World War. Factories and mills produced goods that were bought by Americans and consumers across the globe. It was not alone. During the golden age of capitalism in the 1950s and 1960s, industrial towns like Cleveland in America shared a kinship with others across Europe. They manufactured the consumer goods, from cars to washing machines to clothes, which satisfied the demand of the era. It meant plentiful jobs and security, which at the time that must have felt like it would go on forever.

But it did not. Cleveland became part of what is known as the American Rust Belt. By the 1970s industrial towns and cities in the West declined. Deindustrialisation set in as manufacturing waned and factories were abandoned. Advanced economies, led by the United States, built on the information technologies emerging and evolving, and created service sectors with finance at its heart. Eventually this would power growth, so that by the mid-1990s an economic transformation had taken place. But this was little comfort to those who had relied on old-fashioned work in manufacturing and steel production.

In 1978 Cleveland became the first US city since the Great Depression of the 1930s to default on its debt, and today across the metropolis there are nearly 200 acres of industrial land no longer being used.

By this time a subsidiary of Woolworths, the Richman Brothers factory finally closed its doors in 1992, when its chain of outlets finally went to the wall. Frances Trachter, a spokeswoman, told the press that the business was now 'underproductive and not achieving the kind of returns dictated by the company's financial goals'.[22] Since then, its contents stripped, the great edifice has sat as a silent and empty reminder of Cleveland's proud manufacturing past.

IV

Manufacturing technologies have of course improved massively over the decades. Production today is so much less labour-intensive than even when Richman Brothers was at its height. But automation has not been primarily responsible for the deterioration of these once proud industrial towns and cities. There have not been hundreds of Peppers clocking on. And Western consumerism has hardly abated. The difference now is that we buy so many of our products from overseas.

It is unsurprising that this phenomenon has become an emotive political issue. For the populist left, who favour the Universal Basic Income, it is an indictment of the neo-liberalism which took hold in the 1980s; for the populist right it is an opportunity to attack cheap, global competition, the export of jobs and immigrant labour. And it is economies such as China, emerging so powerfully in the late twentieth and early part of this century, which was to become known as the 'workshop of the world'. It was able to do this because it has been so much cheaper to produce goods there than in the expensive United States or Europe.

Workers at Richman Brothers enjoyed unparalleled working conditions including a 36-hour work week, a fortnight's paid vacation, paid maternity leave and even stock options. Not so for workers in countries just beginning to industrialise. The West simply could not compete with the ultra-low labour costs of emerging economies and meanwhile benefitted from cheaper goods which they imported in abundance.

Simplistic calls to 'bring back manufacturing' are easy. But doing it is much harder, especially as the constituency to which populist slogans are aimed is the unskilled. That is people who would have once enjoyed the security of well-paid work, making hats and coats for Richman Brothers, but who now find themselves as the precariat, reliant on low-paid and uncertain jobs.[23] This is surely the hardest part of manufacturing to 'bring back', even while labour costs in the cheapest economies rise and working standards improve. For while developed economies continue to manufacture, it is the higher end, more valuable, specialist design work which tends to be located there. Ironically while manufacturing as a share of GDP has declined rapidly in the West, actual manufacturing output is surprisingly steady. However, that output simply does not need the volume of workers that it once did. That is, at this end, we can produce more with fewer people.

Naturally this raises questions about the role of the state, how much it should intervene to support struggling industry, to subsidise jobs and to protect products from competition. It raises questions about the balance between the protection of standards and the ability to compete globally. You will not find a national election anywhere in the West where these issues are not raised – though you will also struggle to find them addressed.

But surely these are the arguments of yesterday. And the political debate is all so often stuck in the lost battles of a previous generation. At this inflection point we can look forward to a world that is very different.

On the one hand, manufacturing technology has come a long way since Henry Ford's production line in the early twentieth century. Automation has reduced the need for human hands and performs countless routine and repetitive tasks. But on the other hand, manufacturing really has not been revolutionised for decades. It has relied on cheap labour to increase productivity. It has relied on globalisation. That revolution is overdue and manufacturing technologies are rapidly becoming capable of performing non-repetitive tasks. 3D printing technologies offer endless possibilities. Warehouse robots employing sophisticated AI are becoming competent enough to recognize different objects, sizes and fragility; hold, grip and handle them with the dexterity of human hands.[24] This all means that customisation on a mass scale as the future of manufacturing. Cognitive manufacturing offers the prospect of technology learning from big data to improve quality, foresee production breakdowns, increase productive performance, identify defects, and act to fix them with little or no human intervention. It is the same kind of 'artificial intelligence' used by Amazon to scrutinise job applicants or by Joseph Franklin to prevent suicides, but here deployed across a manufacturing plant to learn, scrutinise and improve outputs at a rate so much more rapid than could ever be achieved by human workers.

But this all means something else too. This revolution does not rely on cheap labour or indeed very much labour at all. Researchers at Oxford Economics analysed the growth of robots. They calculated that each robot introduced in manufacturing can replace 1.6 human jobs. They forecast that as early as 2030 as many as 20 million workers globally will be displaced by these industrial robots, around 8.5% of all jobs. And such is the productivity dividend of using robots, they estimate an increase of installations of 30% would raise global growth by 5.3% or $5 trillion.[25] To put that in perspective, it is something like the GDP of Japan.

Today, a third of all new robots are being installed in China, and the estimate is that 14 million of the 20 million robots by 2030 will be Chinese. The explanation is the incremental nature of their introduction – they are going into the vast factories that already operate in the workshop of the world. But that does not have to be the case. Automation, AI and robotics mean that the location of the manufacturing plant would no longer depend on the supply of low cost workers to remain competitive. For once, this new world could well 'bring back manufacturing' to the developed economies where industrialisation first took hold.

This is a world which has the potential to breathe life back into the dereliction of the Richman Brothers building on East 55th Street in Cleveland. But is a world which cannot bring back the thousands of low-skilled jobs that it once sustained.

V

Cleveland was home to a Hooverville. These were the shanty towns which grew up around the United States during the 1930s to house those Americans who had lost everything in the Great Depression and now moved to where there was work, any work at all. Makeshift shelters housed families in vast, untidy and uncomfortable surroundings. They were named, mockingly of course, after the President, Herbert Hoover, whose time in office was dominated by economic decline and who was beaten by Franklin Roosevelt in the 1932 election primarily on the issue.

The site of destitute migrants – camped around areas where word had got out that work was available – must have been astonishing. Desperate for any opportunities to earn a few dollars, there was often little by way of food. Shelter for most would have been a draughty tent fashioned from tarpaulin, with their few possessions packed away in suitcases.

For those millions of Americans and millions of other people around the world, out of work, mostly unskilled, the 1930s was a decade of great degradation. The great depression brought the boom of the 1920s to a shuddering halt.

Economists have been debating the root causes of the Great Depression since the 1930s, and there are many competing theories. The debate continues to rage partly because the evidence is imperfect, and our perspective is skewed by the massively interventionist policy prescriptions put in place by President Roosevelt as much as the indicators leading up to downturn. This perspective is further skewed by which side of the ideological divide we fall in terms of the neo-liberal experiment of the 1980s. It is skewed once more by the experience of the global economic crisis which started in 2008. So there is no definitive explanation and it is not the intention of this book to referee.

There is one aspect of this episode that could illuminate, however. The 1920s economy was fuelled by consumer credit, as families stretched themselves to buy the latest mod cons. Mass production of that era made new products available but, economically, disproportionally benefitted the small number of factory owners who grew wealthy. This in contrast to the workers who made the goods; the very workers who represented the experience of the vast majority of Americans. You can see the problem here. So unequal was the economy that only those who owned the means of production could exert financial power; and they largely invested profits into the stock market, which we now know was creating a 'bubble'.[26] And so, once ordinary people reached the end of their credit, the economy was failing to empower consumers. That is, there were simply not enough people with enough disposable income to buy all the products the economy was able to manufacture. As a result, goods started to pile up unsold, filling warehouses up and down the country. Productivity growth had been strong in the decade before recession, but wage growth had not kept up. Production slowed and in some cases stopped, putting millions of people out of work and in doing so reducing further the supply of consumers who might buy the products now stacked high and shut away.

Richman Brothers responded to the Great Depression in a very progressive way. The eponymous brothers Nathan, Charles and Henry gave up their pay permanently in 1932. They established the Richman Foundation to offer interest-free loans and grants to struggling workers. The company continued to pay above the Code of the ill-fated National Industrial Recovery Act. But this was far from usual, and the 1930s was a grave time for millions of Americans and Europeans.

The 1920s also felt the tail end of the second industrial revolution, where electrification, mass production techniques and more efficient transportation gave a shot in the arm to productivity. Workers found themselves being replaced by the technology, and unemployment crept up from the mid-20s. Once again this meant overproduction and underconsumption.

While manufacturing was suffering, farm prices fell as a result of expansion, mechanisation (the new farming technology itself largely bought on credit) and improved fertilisers and seed. This in turn led to an overproduction, driving down prices, forcing farms out of business and rural banks to fail.

A dystopian picture that could be painted about the Fourth Industrial Revolution might draw parallels with the depression. It might see a world in which technology can manufacture countless products, efficiently and cheaply, all in factories manned by robots and employing cognitive manufacturing, 3D printing and powerful bioscience. They will not need significant workforces. These factories, like before, would be owned by a small number of people, the problem being once again an overproduction coupled with underconsumption. The Oxford Economics report highlights the real dilemma that new technology presents: robots enable economic growth but they exacerbate income inequality.[27] Already we can see that productivity growth has decoupled with wage growth. That is, while once wages increased with output, pay has stalled while our efficiency continues to accelerate.

Just like with the Enclosure Acts, there is a picture to be painted where there is a replacement of one kind of life with another. Taken away. A picture where jobs are sold in exchange for technology. Does this all sound like a need for 'a compensation in part, for the loss of his or her natural inheritance'?

VI

At the age of thirty-three, Bill Gates was identified by *Forbes* magazine as the world's youngest self-made billionaire. The founder of Microsoft, Gates later held the title of richest man on the planet for thirteen years and spawned a business empire which reached into every corner of that world, practically every business, every public organisation and every home. He has been an extraordinary technologist and visionary who grasped the potential of computer software like few others. Gates has also been a significant philanthropist, pledging to give away half of his wealth with his Foundation, the world's largest private charity. But he is a true capitalist. His personal investment firm, Cascade, holds significant stakes in numerous companies and he owns a property portfolio headed by a 6,000 square-metre mansion set on Lake Washington with its own beach containing sand imported from the Caribbean.

Capitalist or not, in a 2017 interview with *Quartz* magazine he proffered an extraordinary view. 'Right now, the human worker who does, say, $50,000 worth of work in a factory, that income is taxed and you get income tax, social security tax, all those things,' he said. 'If a robot comes in to do the same thing, you'd think that we'd tax the robot at a similar level.'[28] Gates, the one-time most successful capitalist on the planet, the developer who was so instrumental in pushing the boundaries of computer technology, this very same man was arguing for government intervention to slow down automation responsible for replacing jobs. And the mechanism would be the blunt instrument of taxation, which, he believed, would replace the lost income tax which would have been paid by human workers.

It's an attractive, even populist idea. And in echoing the concerns of past technological revolutions, which threatened jobs and livelihoods, throws up similar questions – if not answers.

One big question is how should we tax the means of production if it is not human and the ownership is concentrated in few hands? The dystopian scenario which draws parallels with the Great Depression is that the vast inequality this would create could have a corresponding impact on demand. That is, there would be too few of us out there with the disposable income necessary to sustain production volumes which these new robot workers are capable of maintaining.

Here then is both a moral and a practical case for the Universal Basic Income. Morally it would be about compensating the 'natural inheritance' of those who find themselves out of work, in a way that Thomas Paine would be familiar with, and in practical terms it would be about creating (or at least maintaining) the consumers of tomorrow necessary to sustain economic growth.

It is not exactly a socialist argument, but then again Bill Gates is hardly a socialist. In this new world there could be an economic change so profound that it requires ideas once dismissed as absurd to be re-evaluated. It might be that just how market mechanisms work needs to be supported differently. It might be that a universal income is a capitalist intervention in the Fourth Industrial Revolution.

But it is not 'universal' is it? Taxation is levied at a national level not international. There are some exceptions of course. Blocks like the European Union have some standardised levels of taxation and there are trade agreements which agree tariffs or limit subsidies. But personal taxation and business taxation remain the preserve of the state. As does public spending. You see the problem here? In order to create demand, countries will tax robots and redistribute, but that will make that country's robots more expensive to operate than their competitors.

Manufacturers will simply locate their plants in the places with the lowest robot income tax. And that means that politicians who promise to 'bring back manufacturing' have just as impossible a job as they have today. That is, manufacturing will remain firmly in China, as will everything else.

VII

Before we get carried away, however, it's worth remembering that we have been here before, or sort of. This is the future anticipated by visitors to the 1939 World's Fair in Flushings Meadows. It is the world people had been waiting for through the horrors of war. It is the world that the great economist John Maynard Keynes sketched out in a futuristic paper a few years earlier. And it is the world which seemed to beckon in the optimism of the golden age of capitalism when Richman Brothers flourished and became the biggest clothing manufacturer in the world.

The Associated Press published a 1950 feature, 'How Experts Think We'll Live in 2000 AD', where specialists confidently foresaw the future. And what would that destiny be? Well gone was the pessimism of the past, the fear that technology would make humanity obsolete, and in its place an optimistic vision that the economy would be working for us, providing us with prosperity and huge amounts of leisure. That is interesting, given that memories of the 1930s were still so raw. But it showed the optimism of the era nonetheless (even if it got most things quite wrong).

Part of this would be because of domestic technology and vastly improved public health. They predicted, for instance, that 'a housewife may use an electronic stove and prepare roast beef in less time than it takes to set the table'. Flying 'will be accepted by the public as readily as mid-century's automobile and train... we shall be neighbors of everyone else on earth.' These innovations were not far from our reality if we think of microwave ovens or budget airlines. But fundamentally, it foresaw an economy which 'by the end of the century many government plans now avoided as forms of socialism will be accepted as commonplace. Who in 1900 thought that by mid-century there would be government regulated pensions and a work week limited to 40 hours? A minimum wage, child labor curbs and unemployment compensation? So tell your children not to be surprised if the year 2000 finds a 35 or even a 20 hour week fixed by law.'[29] It is an interesting idea that more successful capitalism leads to 'socialist' policy, but what is true is that social progress has followed industrialisation and economic growth the world over. That said, it is amazing how long this vision that technology would replace jobs persisted. As late as 1966, the pioneering pollster Bud Lewis told his radio listeners that 'we're going to have to readjust our old, puritan perhaps, concepts of what a person should do with his life... there's not going to be all the jobs there used to be around.'

Of course, that did not happen. Leisure time enjoyed in the West might have increased a tad, but this is not because humans have too little to do. The twenty-first century saw more jobs created than ever before. As anyone who has to manage a work email will attest, technology has given us more to do, not lightened the load. And where jobs have become obsolete because of technology such as switchboard operators or door-to-door salesmen (especially of encyclopedias), the result has been freeing up productive capacity to do new things – just as it did during the first industrial revolution.

Is anything different this time? Keynes set his new future in 2030, in touching distance of where we are today. He saw 'technological unemployment' where that technology advances faster than new labour opportunities and in doing so solves the 'economic problem' of producing sufficient amounts to satisfy our needs or wants. Are we really closer to the fantastical futurology Keynes envisaged? Is Bill Gates right? In part, at least, he might be.

VIII

The Council of the Great City Schools 61st Annual Fall Conference in 2017 was held barely a fifteen-minute walk away from the decaying Richman Brothers factory in Cleveland, though few of those who attended gave it much thought. The agenda was focused on the serious topic of 'advancing the state of urban education'. And this five-day gathering attracted delegates who might be urban school superintendents, college deans and school board members.

The star speaker that year was none other than Bill Gates. The Gates Foundation has invested many millions of dollars into improving education and skills since it became involved in the cause in the year 2000. The guiding principle was to ensure that public education prepared high-school graduates with the skills to succeed in the workplace and has sponsored numerous pilots to evidence the very best ideas.

Gates told the conference:

'We are ... interested in what role we can play to prepare students for the dramatic changes underway in the workforce. We have to make work-related experiences a consistent part of high schools in ways that build student engagement and relevant skills, and that put young people on a path to credentials with labor market value in our future economy.'[30]

Nonetheless he acknowledged just how far short US public education was falling. Aggregate numbers, he said, mask the reality that while white American students achieve outcomes comparable to the best in the world, the likes of Finland and Korea, schools were failing black and Latino children, whose scores were among the lowest performing OECD countries like Chile and Greece. He went on:

'Without success in college or career preparation programs, students will have limited economic mobility and fewer opportunities throughout their lives. This threatens not only their economic future but the economic future and competitiveness of the United States.'[31]

That great challenge for schools in deindustrialised places like Cleveland is the nature of the skills the present economy (let alone the future economy) demands. These are places where employers like Richman Brothers of well-paid, low-skilled work are no longer the norm. And this makes Gates's earlier argument that we should tax robots somewhat incongruous.

Imagine if governments had intervened at previous inflection points to prevent change. The truth is that it might have slowed down the progress which eventually created all those jobs and which in turn led to social change and the protections enjoyed today.

While superficially attractive, there is a real question as to whether we really want to tax progress. Do we really want to slow down the pace of advancement in order to hold on to low-skilled jobs for longer? Looking back now, should we have avoided electrification in order to keep the lamplighters in work or delay development of the microcomputer to prevent the human office 'computer' from becoming obsolete? Of course not.

The less emotive case against taxing technological progress is linked to the speech that Gates gave to the Cleveland conference. Schools are already falling short in producing young adults with the skills demanded by the economy today, let alone the one when robots take over. Read virtually any employer survey in the developed economies and you will find one very clear message – there is a shortage of skills already.

Let's take the economy in 2019, before the coronavirus crisis hit. According to the Bureau of Labor Statistics itself there were the best part of 7 million jobs unfilled in the US economy.[32] That is a lot of posts, meaning a lot of potential productivity. One of the main reasons that employers cannot recruit to these positions is that there are insufficient candidates applying who have the requisite skills. And in a world that is rapidly being driven by fantastic technological advances, it is astonishing to note that the skills shortage is not simply about science and technology.

Take a look at what US-based SHRM – Society for Human Resource Management – says. According to their survey, three quarters of members who have difficulty recruiting blame a skills shortage. Technical skills are in short supply, but so are the soft skills of problem solving, critical thinking, creativity and innovation; ability to deal with complexity and ambiguity; communication. And they blame the education system for failing to address these shortages.[33] The story is the same in practically any skills survey you might find, practically anywhere in the developed world.

And we can understand how we got here through the work of Claudia Goldwin and Lawrence Katz, who contend that over the last century or more there has been a race between technology and education and inequality in the US economy. And whereas once there was a growth in education which drove American economic pre-eminence, recent decades have seen an educational slow-down and with it increased inequality.[34]

We are then in a world where those high human skills are in great demand: creativity, innovation, collaboration, adaptability, cultural intelligence and leadership. But as Bill Gates was at pains to point out, the education system is producing too few graduates with these attributes.

IX

Some 26,000 people work for the United Overseas Bank (UOB). Headquartered in modernist twin towers, the tallest of which stretch 280 meters into Singapore's skyline from its base in the historic Downtown Core area. A global bank, UOB services Asia Pacific, Europe and North America through 500 offices. It offers a range of professional services including private banking, asset management, venture capital and insurance.

This is an organisation which takes its talent seriously. It recruits high-calibre graduates and it continues to train them. In its 2017 Annual Report, it boasted of the $21.3 million spent developing people across the group. Meanwhile, its eighteen-month Management Associate Programme tells the story of the sort of people who work there. The objective is 'to attract high-calibre, early-career talents from across the region to take on roles in UOB'. In 2017 a staggering 22,000 young men and women applied for this programme and a fortunate – and talented – thirty-two 'fresh' graduates were selected and put through their paces.[35]

You will get the sense then that the sort of people recruited by UOB are not the kind of high-school graduates that Bill Gates had in mind when he spoke to the Council of the Great City Schools in Cleveland. Gates is concerned with those who have failed to develop the basic skills needed in today's disruptive economy. UOB, by contrast, has its pick of the brightest, most highly skilled university graduates in the region and the world. To be selected as one of those thirty-two is a massive achievement. After all, this is a huge competitor pool comprising similarly well-qualified people. To be one of the chosen few requires a candidate to display aptitude and skill that really marks them out from other very capable individuals.

Because of this talent it is all the more alarming when an organisation like United Overseas Bank allows the world to peek inside and share its anxieties. The Bank takes its people development seriously. And in 2019 it announced a major intervention to make sure employees 'remain relevant'. Think about that for a moment. An organisation which is able to recruit some of the most talented graduates in the world, one which invests tens of millions of dollars in training its staff, a bank whose expert and qualified employees delivers professional financial services. This organisation is worried those people might not have the skills to stay relevant at this inflection point.

In 2019 UOB rolled out a programme called 'Better U' – right across the group. This twelve-week foundation course focuses on five 'essential' competences: encouraging a growth mindset, developing complex problem-solving skills, acquiring skills in the fields of digital innovation, human-centred design and data. More than two thirds of employees were to complete Better U by 2020. Just look at those competencies carefully. They do not need vast amounts of knowledge but they do require the active development of higher human skills and that makes it different from passive corporate programmes of old, which loaded staff up with specialist expert know-how and taught them how to do things. 'Encouraging a growth mindset' is of a different order and signals a significant shift away from the old mindset – the industrial mindset – located as it was in industrialisation and the organisational culture and professions which we have lived with since.

The head of UOB group human resources, Dean Tong, explained the motivation: 'To help our people find their place in an environment where work is being transformed by technology and customer expectations,' he said, 'we designed the Better U programme to nurture a mindset that is curious and open to new avenues of personal and professional growth.'

But is this putting technology at the centre rather than people?

Not only is there a shortage of skills in today's economy, but the truth is that in tomorrow's world the new technology poses the greatest threat not to factory workers but to the professions. Liisa Koskinen is doubly fortunate. Fortunate to receive a no-strings universal basic income payment from the Finnish State but perhaps more fortunate to have been an accountant since qualifying in 1988. This has kept her in permanent and well-paid employment for more than three decades. Like being a lawyer or teacher or doctor, an accountant is one of the professions. To be a professional is to be part of a club. And any club worth belonging to has barriers to entry. To be a professional is to hold recognised qualifications, to hold specialist knowledge, to behave in accordance with the rules of a regulatory body which decides who joins. Because of this specialist knowledge and the fact that not just anybody can call themselves an accountant, a lawyer or a doctor, comes the privilege of being able to charge a substantial fee for the work. And it has been this way since the nineteenth century when the professions really established themselves in the industrialised world.

Having those credentials means that rather than being determined by the market, those professional services are determined by expertise and commonly held standards.[36] That is professions are associated with the development of an industrialised, capitalist economy, but by their very nature limit competition. After all, it is not possible for just anyone to decide to practice medicine or the law. A professional enjoys a particularly exclusive form of human capital that can only be replaced by another similarly expensive expert. But that privilege might not be as permanent as it seems.

It is interesting to think about the value of knowledge, because that is what has underpinned the dominance of the professions. But today knowledge is no longer exclusive. And it is increasingly free. Most of us carry around a piece of technology in our pockets or purses that can discover in seconds virtually any piece of knowledge. In that sense the smart phone and the internet have democratised knowledge. They have turned knowledge from being the preserve of the specialist into a freely available commodity. And the trajectory of machine intelligence suggest that it is at an inflection point where it develops from providing knowledge to interpreting that knowledge. What does this mean for professionals whose occupation is to a large degree dependent on expert knowledge?

Expert work, let us remember, has been made redundant in the past. Richard Arkwright is commonly considered to be the pioneer of the factory. In 1769 he and John Kay patented their invention – the spinning frame. Arkwright soon opened a factory, probably the first of the industrial revolution, in Cromford in England's Derbyshire. And within twenty years the factory employed some 800 workers churning out vast quantities of yarns with really very little expertise needed. That technology was not producing something which had not existed previously, but it did replace the expertise of individual craftsmen whose skill was uncommon. An early story of industrialisation is this very replacement of specialist craft skills with machinery which can be operated by fewer humans with lower skills. The tragedy too is that Edmund Cartwight's power loom which emerged towards the end of the century put well-paid weavers out of work altogether. Without demand for their skills, many craftsmen accepted the offer of low-paid work in the very factories which had replaced their expertise.

Today it is not manual dexterity but applied knowledge which is threatened with extinction. The professions which have held on for so long to their status look to a future where that expertise could be performed by an 'intelligent' machine. Is it inevitable that professionals will suffer the same fate as the craftsmen, that technology will mean that expertise can be delivered by someone who is much less skilled (or without human delivery at all)? It is not too fanciful to imagine a machine that can diagnose and refer more effectively than a general practitioner or one that can advise and complete contracts for most day-to-day conveyancing. Organised by the American Society of Neuroradiology and the Medical Image and Computer-Assisted Interventions Society, the Brain Tumor Challenge tells us how good technology already is. In 2021, a diagnosis competition between doctors and an AI system saw humans achieve 66%

accuracy in thirty minutes when diagnosing 225 cases. The AI was right 87% of the time and took just fifteen minutes. But perhaps doctors and lawyers do not need to worry about their new role as low-paid machine operators... If the professions adapt to the opportunities of the new world.

The democratisation of knowledge and the emergence of technology that can perform evaluative functions with that knowledge does not mean we no longer need expertise. In fact, it could increase the demand for expertise. But perhaps no longer the sort of expert which has characterised so many in the professions. After all, the sort of expert we pay to advise or diagnose routine matters, the sort of experts who, presented with the same problem, can be relied upon to come up with the same answer, this is work that can be performed by an intelligent machine. Experts are needed to make sense of knowledge, to make sense of the changing world, to create new knowledge and to innovate. We need specialists more than ever who will work with technology and with each other.

There is a lesson to be relearned from the inflection point of the first industrial revolution when members of the Lunar Society met. They brought specialists together to create, to innovate and to wonder. As we move from the old industrial mindset to the new growth mindset, that wonder needs to be rediscovered. You will see that this is the very journey upon which UOB has embarked. There needs to be far greater fluidity across disciplines and opportunities to embrace the opportunities that the new technology offers.

This is why the most forward-thinking organisations are planning for the workspace of the future. They are thinking beyond today's open-plan offices with their breakout pods and meeting rooms. They are thinking about how experts come together and how technology supports them. The partnership between office specialists Steelcage and Microsoft is a great example of this. It comes from, as they put it, a 'shared commitment to put people at the center of how place and technology intersect and empower individuals and teams to do their best work'. And the results are some fabulous places to work. Places we would all like to work.[37] They include 'Maker Commons', where ideas can be thrown around and then put into action; the 'Ideation Hub' for mixing ideas with remote teams; the 'Focus Studio' for time alone.

This might be the best way for technology to interact with people in the workplace, but is it how people best interact with each other? Should we be thinking about putting technology at the centre of what we are doing, or should we be putting people there? And if the post-profession society is upon us, is this the best way for specialists to create and innovate?

XI

Tatty, dilapidated, cold, leaky and constructed of plywood, it is extraordinary to think that a temporary structure put up hastily during the Second World War ended up as home to some amazing things. And amazing is not an exaggeration. Consider hosting no less than nine Nobel Prize-winning scientists, seeing the creation of the world's first atomic clock, the construction of an early particle accelerator and even the birth of modern linguistics.[38] These things and more could be found in the crumbling walls of what turned out to be a miraculous work environment.

'Building 20',[39] as it was known, was only three floors high but stretched untidily down Vassar Street on the campus of the Massachusetts Institute of Technology (MIT). Offering 23,000 square metres of workspace, it stood on a basic concrete base, the frame covered with unattractive asbestos cement board. Intended to be pulled down soon after the war, no one really worried that the windows didn't fit properly or that wires hung down from the ceilings or that pipes were exposed or that it was near impossible to find your way from one part of the structure to another. This meant that if you were trying to find someone or something, you often stumbled through another occupant's workspace, interrupting whatever they were doing. And the fact that it was a condemned building meant that when it did survive the end of war, because MIT was desperately short of space, those occupants felt little obligation to look after it.

Those occupants were from all over the university. The campus was overcrowded and academics of all kinds were housed there, taking over the cramped rooms which were connected by long, taciturn, corridors. It was never a long-term plan and so little thought was given to how they were organised. Those occupants, however, achieved some extraordinary things over the five decades until the structure was eventually razed to the ground in 1998.

Did they do this despite the ramshackle accommodation or because of it?

Clearly there were some brilliant people working in Vassar Street during those fifty years. These are people, whether Noam Chomsky or Amar Bose, who were destined to achieve great things. But institutions can all so often stifle the emergence of great ideas. Most accounts of Building 20 (and there are lots) tell the same story. The reason that so many ground-breaking developments – from the redevelopment of hi-fi speakers to the improved understanding of the nervous system to the invention of single antenna radar to the development of hacker culture – owed so much to the decrepit nature of that building.

Just think of the possibilities when thinkers and experts of all different types are thrown together to use a space without planning or even artificial organisational silos. Just think of the new ideas that could be generated through random exchanges as one expert meets another in a different discipline, passing in a corridor, stumbling through their lab or over a drink at one of the many parties. Just think of the freedom that comes with working in a building that no one cares about. If a wall is in the way when an experiment is being conducted, the answer is simple: knock a hole in it!

Building 20 acquired a legendary reputation as one of the most innovative and creative places on earth. It brought together experts without disciplinary demarcations and incubated their interactions. While many have tried desperately to recreate the magic of Vassar Street, few would seriously contemplate downgrading modern, safe and clean workspaces. But the thirst for understanding that it created is surely closer to the Lunar Society than these attractive tech-centred spaces and serves as a future lesson for success, freedom of thought and action. Crashing human specialisms, expertise and disciplines together and making use of new technologies has such exciting possibilities.

XII

Those possibilities are desperately needed because the world faces such great challenges on the inflection point of the Fourth Industrial Revolution. We all know what they are:

climate change, sustainability, terrorism, conflict, disease and poverty. These are some of the great global challenges that demand the collective might of the world's specialists and all technological possibilities imaginable. And soon! There are those who doubt this is possible in a global economy dominated by capitalism and the profit motive. But should we be more optimistic?

Remember William Fothergill Cooke who, in the mid-nineteenth century, created an electric telegraph system? Well, we might consider just what motivated him to push forward with his invention and make it such a commercial success. And what was his pitch to financier John Lewis Ricardo? Was it all about how much money the two of them were going to make?

Well if that were the sole or even the main motivation, both men were to be sorely disappointed. In his history of the electric telegraph, Jeffrey Kieve highlights just how unprofitable this venture initially was.[40] It was a major endeavour into new (disruptive) technology which required the foundation of a brand-new market that soon attracted competition. Despite risking considerable capital, Ricardo's shares were virtually worthless in the firm's early years. But what they created was the world's first public telegraph company. And 'created' is the important word here.

There is a very good reason that rational explanations of market capitalism have been reappraised in recent years and it is not only because human beings' economic behaviour is so often far from rational. The long-held definition of economics is about allocating scarce resources. It is 'the economic problem' as Keynes put it. Meanwhile the neo-liberal approach to economics, built on the work of Milton Friedman, and which took hold in the 1980s, assumes that the scarcest of resources is capital. Why else would maximising shareholder value (profits) be elevated to the prime objective of the system?[41]

Of course, there needed to be an economic system which incentivised Ricardo to risk his capital. That capital was indeed scarce and could have financed alternative ventures. It is the same principle as the Enclosure Acts, which recognised land capital as a scarce economic resource. But more than that, there also needed to be a system where capitalism could solve society's problems. Think about it. When looked at in this way, there was something which was much scarcer in this story than Ricardo's capital: it was the creativity of Cooke. You see, the purpose of Cooke's invention was not to maximise Ricardo's returns. No, it was to solve a problem faced by society. And Cooke was not allocating resources better, he was creating new resources. The reason we are still talking about the electric telegraph is not the attraction of capital (though this was necessary), but the allocation of creativity.

The great value of the capitalism to have emerged from industrialisation is not the profits derived from efficiently allocating the world's resources in some linear fashion. It is a much more dynamic network – spread across cities, countries and the world – which facilitates creativity, innovation and addressing the problems of society.

From Richard Arkwright's spinning frame at the beginning of the industrial revolution, to electricity and steam of the second, the hats and coats produced by Richman Brothers, the computer technology of the third revolution and what Bill Gates was able to do with software, the great advancements made in the shabby surroundings of Building 20, the complex financial services offered by UOB, Pepper the robot and the emergence of machine intelligence. All of these things and everything in between, from antibiotics to the motor car to flight to satellites to the internet and the mobile phone – you name it, these are about providing solutions to society's problems. And as a result, human beings today are healthier, wealthier and have so many more opportunities.

One of today's great challenges is tackling inequality, and Industry 4.0 risks making this wider not narrower. Industrialisation has meant social change, healthcare, citizens' rights, longevity, wealth… freedom. And it is no surprise that the great enrichment was about free thinking and fresh ideas and creativity. Just as in the Fourth Industrial Revolution.

Naturally, this is not all about private enterprise. Democratic government has had a huge role to play in facilitating and shaping. The internet itself is a product of the public sphere, providing capital at the earliest and riskiest stage of the innovation cycle. And research demonstrates that, contrary to popular opinion, the public sector is just as good at innovation as the private.[42] But governments have to trade freedoms with protections and protections with growth.

The Electric Telegraph Company is an intriguing example of this. Remember that as the Chartists gathered in London's Kennington in 1848 to demand votes for all men, the British government feared insurrection. It used its reserve powers to take control of Cooke and Ricardo's company, disrupting the communication of the well-organised Chartists.

The difficulty today is that the biggest problem the planet has arguably ever faced is a direct product of industrialisation. And the trade-offs that government have had to make have tended to side with economic progress rather than ecology. For capitalism to solve this societal problem will need the allocation of capital, but it will also require the deployment of even more powerful resources. And as we look beyond this inflection point, the challenge is already here.

The biggest scarce resource of the future is a scarcity of now: it is creativity, it is adaptability, it is innovation, it is leadership. It is the most human of skills.

Perhaps the big story of these debates is ultimately a positive one about great problems being tackled by ingenuity and progress. Or maybe there is a more sobering analysis that points to these great problems identified here as not only existing issues, but ones which, despite the economic growth, the great enrichment, the technology, the social change, remain debates we have failed to properly address. And as we peer over the inflection point to the exciting world of tomorrow, they remain stubbornly with us.

The question is left hanging: how do we ensure that all have a real stake in society? It is confusing of course it is. If it were not so, we wouldn't need the debate.

Why Disruption Is Inevitable and Why Everyone Matters

'At first, we just worried about losing points, but now we got used to it.'

I

Imagine if the machine learning developed by Amazon or Pepper the robot were being used for a different purpose. Imagine if, instead of rejecting job applicants or helping the elderly, it were employed by the military, by law enforcement or by private security.

Imagine it had the capability to kill.

Such a prospect takes us into a whole new world: a new world of responsibility, of ethics, of behaviour and of society.

Actually, perhaps we should not be too alarmist. The emergence of (narrow) Artificial Intelligence in areas where they overlap with risk of death have invariably preserved life. Driverless cars, for instance, promise to be much safer than human beings behind the wheel. Removing human error from long-distance trucking or piloting aeroplanes or handling heavy machinery or diagnosing cancer or performing routine surgery offers the same advantages.

But the prospect remains. As machines become more intelligent, they will make more decisions that we would hold deep down to be innately human. The sort of moral decisions steeped in personal values and philosophy. The sort that need chewing over before and after the fact. This involves not only what one might think of as defensive decisions – say an automated car mitigating loss of life in a road accident – but also offensive ones – an armed drone being sent into battle in a foreign city.

And as machines learn – just as Amazon's recruitment system did – they will eventually take these decisions for themselves; going far beyond following programmed logarithms. They will decide who lives and who dies when defending life, and they will decide when to pull the trigger when in offence.

Aside from the question of where responsibility lies for the actions of these machines, whether that sits with the state or formal organisations or private individuals or some other new entity, this prospect raises a new question about what it means to be human and underlines the case made here that this inflection point really does signal 'disruption' to the way we live.

II

This chapter is about the idea of disruption, what it means for us and the extent to which this is an inflection point that marks what can truly be described as a 'revolution'. A revolution would have to mean fundamental change in the way we live, work, make decisions, perceive the new economy and structure of society. And that is big.

Disruption is an overused word at the best of times. With particular focus on technology, disruption is associated with business and the management thinker Clayton Christensen, who coined the term back in 1995.[43] Here, it refers to the ability of smaller businesses with fewer resources to challenge successfully larger established firms by targeting the unmet demands of some of their customers. Having established this foothold, they are then in a position to compete for the wider customer base. They disrupt the status quo with their innovation, fresh business models and technology, and this can be seen time and time again. But such is the interconnectedness and volatility of the times in which we are living that the term is also applied more broadly to the very things we experience around us. Debates in the World Economic Forum refer to our 'age of disruption' and mean that the uptake of new technologies in particular is rapidly altering everyday life.[44] At the risk of 'getting sloppy with our labels',[45] this book uses the term in its loosest sense to describe fundamental change going on around us not only in business but in the wider economy too, and in the workplace, in society and in our politics.

The question about disruption is not simply about what the future holds but just how do we get there? That was partly the point of the last chapter, because it takes place through age-old debates. This chapter, though, considers where that journey takes us next and the forks in the road ahead for the way we work and live. It connects the idea of disruption with a popular concept associated with a twentieth-century economist called Joseph Schumpeter, and that concept is creative destruction. This chapter considers what Industry 4.0 means for us, as human beings. It remains optimistic but recognises the dark side of what comes next.

III

Think about your job. What do you do each day? How would you describe your work? The chances are you don't think of your job as a simply set of tasks. Most work is far more nuanced than that, and of course it is in that nuance that value is created. And yet when we discuss the growth of technology, whether it be machine learning selecting job applicants or robots in factories, there is a natural tendency to identify tasks which machines can increasingly do: make this, build that, drive, calculate, diagnose.

The last chapter discussed how we are already facing a hollowing out of the economy and a post-profession environment. And tasks are already being assumed by machines. Take Amazon once again and their global network of fulfilment centres. If you were to walk into any one of them, you would be greeted with a similar sight. A vast warehouse with line upon line of storage. The capacity to gather and despatch goods ordered to customers with the ultimate efficiency: speed, accuracy and cost effectiveness. But there is perhaps a singular difference to be found between centres in different parts of the world.

If you were to walk into an Amazon fulfilment centre in, say, France and compare it to the one you visited in India you would notice a subtle contrast: people. In India, the fulfilment would still largely be carried out by human workers taking goods from shelves, packaging and despatching. Labour costs there still, just, make it justifiable to employ people to perform these tasks. But in France, people cost more and that additional cost has made it worthwhile to invest in robots. Here it is machines who are tasked with the job. They are at least as effective and overall cheaper than people. These robots are not intelligent but they are coordinated centrally, meaning their value is greater than the sum of their parts – that is, like people, they know what other 'workers' are doing and they help each other out. The capacity for machines to efficiently and cheaply perform work-based tasks is immense, and many onlookers would characterise this as a glimpse into the future.

But this is unfair to both human *and* machine. The sort of jobs which are set to experience the biggest technological encroachment in the near future are hardly characterised by a set of repetitive, routine tasks. These are jobs which deploy experience and expertise to problems and contexts. Think of the lawyer evaluating a client's case or a general practitioner diagnosing a patient's ailments. And as machines become more intelligent, they are not simply replicating tasks. No, they are assimilating information, they are learning from how things have been done in the past, they are making improvements, they are extending their own capabilities. And what this means is that the possibilities of this new world of technology are not simply about machines replacing human beings but is much more exciting than that. It is about technology that can liberate human potential.

Think about a person doing your job say a century ago, or fifty years or maybe just twenty-five years ago. It is likely to be both recognisable in what is being achieved but maybe rather unrecognisable in just how he or she goes about doing that. Those tasks and functions will have evolved and the technology will have had a fundamental part to play in that evolution. Moreover, the technology which has been adopted is likely to have significantly improved productivity, accuracy and capacity. New types of work created; existing work transformed.

IV

We are at the inflection point of a fundamental change. That much is clear. And the prospect before us is on a scale none of us has experienced in our lifetimes. And just think, there are still those among us who experienced economic depression, world war, the atomic age and the Cold War, the emergence of the microcomputer, the neoliberal experiment, the collapse of communism, globalisation, the banking crash, the Arab Spring, the coronavirus crisis...

But what does an inflection point feel like? How will we experience this particular inflection point? The truth is that we do not really know. That is, the Fourth Industrial Revolution might occur in such increments that we never really notice it happening in the here and now. Instead, one day we might look back and marvel at the change that has taken place over the years. We might reflect on how our lives are so much different than in the past but that the pace of digitalisation and AI was such that it did not unnerve.[46] That must have been what it was like even for those born in the late nineteenth century looking back from, say, the 1950s on a world which had changed from steam and horses to nuclear power and the brink of space flight. It was certainly what it was like for the millions of people who were forced to work from home at a moment's notice when the coronavirus crisis struck in 2020. For many, setting up home offices was relatively easy, with interactions facilitated by widely accessible online video conferencing, file-sharing and powerful mobile devices. Had the crisis happened just fifteen years earlier, it would have found a more analogue world. Back in 2005 those that had the internet were still predominantly dial-up, and mobile communications were in a much more primitive incarnation. It would have been a very different experience and one that would have been much harder for workers to adapt. So it was the incrementalism of the change, even over just fifteen years, that accounts for why we did not quite notice how momentous the progress was. It is another reminder that the Fourth Industrial Revolution will be shaped not only by technology but also by circumstance. Strangely, in retrospect, organisations across the world invested heavily in powerful communications technology but never really bothered to deploy them to their potential until crisis forced their hand. And that could happen again albeit on a far grander scale.

Or it could be quite different.

Imagine if the technology burst upon us in great waves of capability, capacity and coherence. Imagine if the pace of change was near uniform across sectors and functions. This fast, sweeping, technological change is what Bill Gates feared and wanted to slow down when he advocated taxing robots.

Just what the new technology is capable of achieving is the focus of most discussion. From that the implications and possibilities can be surmised. 'What?' is a big question, but so are 'when?' and 'how?'. The faster it happens and more widespread its implementation, the bigger the impact. This means that velocity of change must be thought of as a determinant (if not *the* determinant) of disruption. How quickly can or will technology such as robots displace human workers? How fast will the manufacturing revolution occur? How rapidly will AI become proficient at diagnosing illness, drafting company accounts, giving legal advice or delivering news reports? There is naturally a momentum to this where an initial investment in, say, robotics increases productivity which allows for more investment and displacement of human workers and at the same time cheaper new technology. Of course, technology has replaced people in the past – remember the lamplighters, the office 'computer' or even the automation in car assembly lines. But this has never been a 'big bang'. That is, the burst of new technology did not hit all forms of work at much the same time in a condensed period. That was not even the experience prior to the Great Depression which caught the tail end of the second industrial revolution or with the impact of the microcomputer in the third industrial revolution which meant steady change over many years.

The lessons of the past, nonetheless, centre on the new forms of work which are created through progress, growth and technology. Workers being displaced free up productive capacity to do new things. This has happened time and again despite the apocalyptic warnings. This time *might* be different, but it does not mean new forms of work will not be created or indeed that technology will fill all those vacancies where there is currently a huge skills gap. After all, the demand for these human skills in adaptability, creativity, innovation and leadership, will only expand. These are the scarce resources of today and tomorrow!

Nonetheless, if change happens fast, displacement happens first. This is creative destruction.

V

'Early in life I had three ambitions,' Joseph Schumpeter once said. 'I wanted to be the greatest economist in the world, the greatest horseman in Austria, and the best lover in Vienna. In one of those goals I have failed.' A century on, Schumpeter is one of the most enduring economic thinkers of the early-mid twentieth century. Short, bald and charismatic, he came to prominence in his native Austria before later becoming a fixture at Harvard in the United States. At a time when

Keynes was becoming such a celebrated voice, Schumpeter not only challenged that notoriety but managed to forge a distinctive path: different from the laissez-faire of developed industrialisation and different from the emerging demand led ideas of Keynes and the post-depression economy.

The 'essential fact' he identified about capitalism was what he described as 'creative destruction' and it is a powerful idea. Schumpeter described this as a 'process of industrial muta-tion... that incessantly revolutionizes the economic structure from within, incessantly destroying the old one, incessantly creating a new one'.[47] So it means that creativity and destruc-tion are two sides of the same coin. They need each other. It means that our economy is never static and instead progress in technology, supply chains, processes or even ideas can make businesses or whole industries obsolete. But that very destruction allows for the creativity which builds new busi-nesses, new industries, new ideas and greater productivity. Economic development is not simply bound to experience such destruction, Schumpeter argues, it actively requires it.

The consequence of such ideas is that rather than preventing progress (as the Luddites did when they smashed the looms, or as Kiichi Ishikawa might have been thinking when he damaged Pepper, or Bill Gates appeared to suggest with his proposal to tax robots), we need to accept, embrace even, this creative destruction as necessary to support economic growth and new, different jobs, roles and contributions to our economic societies. And at this inflection point, we need to really understand what it means.

For Schumpeter, the inventor and the entrepreneur are the central figures in this process. The inventiveness and risk taking of these people means a rapidly growing economy whose better, more efficient ways of doing things make old ways obsolete. These are the people who supply that scarce human resource of creativity and are prepared to risk capital. Businesses go bust while others rapidly adopt the new. Jobs are lost. But new ones, better ones, are rapidly generated from the waves of creativity which come in its wake.

Joseph Schumpeter was the economist who really saw that it was innovation that drives our world.

VI

If you were to follow the road inland from UOB's headquarters in Singapore, as many of its employees do regularly, you will come to Bishan-Ang Mo Kio Park. This 153-acre paradise of trees, parkland and the tranquil Kallang River sit beneath the urban sprawl of Singapore with its great blocks stretching upwards to the clouds. The park is where Singaporeans meet, walk, exercise and enjoy nature. And it is a place which faced heavy restrictions when social distancing was required to combat the coronavirus in 2020.

There is an old saying that necessity is the mother of invention. And so if we want to find disruption and innovation, the coronavirus pandemic is surely one of the biggest natural experiments of all time. To address the new demands of crisis, human ingenuity is challenged and new ideas are rapidly formed.

Perhaps the least significant of these is what happened in Bishan-Ang Mo Kio Park. For running among the grassy meadow-like parkland and through the trees of the popular River Plains section could be found a small, yellow, robot 'dog' named Spot. Its four independent legs make it ideal for manoeuvring through the park's varied terrain and over the obstacles it encounters. Fitted with cameras and speakers, Spot enables accurate estimates of the number of visitors to the Park, can broadcast social-distancing messages and reduces the need for both human staff numbers and their own social contact.

While maybe the most trivial example of innovation, when compared to the search for a vaccine, the design of new medical ventilators or the development of antivirals, this little robot is a great example of technological adaptation in response to a disruptive crisis. And the chances are that innovations like these, hastily designed to address the crisis of the moment, will become permanent features of life. It will be of little surprise to many people that the extraordinary experience of the coronavirus was responsible for at least a slight acceleration in technological change.

VII

But hold on, that technological change is already with us. This book has delved into the digital changes, trends and possibilities of the Fourth Industrial Revolution, and it is clear the revolution is happening, the innovation is before us, the transformation is palpable. And it is having an impact on the way organisations are being led, prosper or decline.

The global consulting group McKinsey has taken a close interest in corporate longevity for some time now and their analysis is widely reported. If you go back to the 1930s, they say, and examine the companies listed on the S&P, the average life of a business was ninety years. That life expectancy has steadily fallen since such that by the late 1950s it was sixty-one years, twenty-five years by the 1980s and today less than fourteen years. That is a big drop. And the implications are that three quarters of companies listed on the index in the last decade will not survive the end of this decade. Businesses might merge, be bought out or simply fail. But the inescapable truth is that the big players dominating the stock markets today are those global tech giants like Amazon and Microsoft. These businesses are increasingly venturing into new markets (for instance the $8.45bn acquisition of MGM by Amazon in 2022), while non-tech companies can be seen to be increasingly digital (McDonald's spent £300m on start-up Dynamic Yield in 2019 for its AI technology). Meanwhile, even stalwarts of the past which are still listed, whether they be General Electric or General Motors, do not come close in terms of market capitalisation. And this is perhaps why even today's giants are far from complacent.

The billionaire boss of Amazon, Jeff Bezos, has made a habit out of predicting the inevitable demise of the company he founded. He readily accepts that Amazon 'will be disrupted' and in 2018 told his employees that 'Amazon is not too big to fail. In fact, I predict one day Amazon will fail. Amazon will go bankrupt. If you look at large companies, their lifespans tend to be thirty-plus years, not a hundred-plus years…If we start to focus on ourselves, instead of focusing on our customers, that will be the beginning of the end… We have to try and delay that day for as long as possible.'[48] And that is an insightful view of modern corporate strategy. It is an acceptance that the organisation cannot survive forever, or even for more than two or three decades, and that business leaders simply need to engage in a battle for delayed corporate death.

The conclusion which is so easy to draw from this, whether you are a consultant, manager, academic or CEO of a tech giant, is that businesses which fail to innovative simply cannot survive. And the circumstances are compelling when you think about the McKinsey analysis. In some instances, certainly, there is what Clayton Christensen would call disruption, where smaller innovators attack established markets with cheaper, inferior, products and services. There is much more of what Joseph Schumpeter would recognise as waves of creative destruction, where the demise of old ways of doing things are replaced brutally by new innovations.

But there is something deeply unsatisfactory about this narrative which, to be sure, is driving so many corporate strategies around the world. What is worrying is the prevalence of a kind of self-determinism in the approach. That is, there is such acceptance of the short-term that corporations become no longer prepared to even properly consider the longer-term. That the inevitability of corporate demise is embraced almost as a strategic destination. And that the pressure to innovate, in whatever form, is so great that it takes over as the mission of the organisation.

Take this compelling case, published during the coronavirus pandemic by Deloitte, titled 'Swim, Not Just Float'. In it, the authors acknowledge the global unpredictability of the pandemic and how this leads many organisations to seek the comfort and protection of the status quo but argue that 'this approach can leave companies exposed to the risk of being disrupted, as those companies that are lagging in their digital journeys may be more likely to fall prey to competitive pressures. It is thus more important now than ever before for companies to focus on innovation in existing/new products and services and build new business models to enable them to thrive in the fast-evolving economic environment.'[49] That is the narrative, the prevalent conventional wisdom, that organisations must disrupt or be disrupted, or at the very least protect themselves through changing. Organisations must be adopters of new technologies or suffer at the hands of competitors which have the advantage. They must innovate or die.

We are at the inflection point of a digital revolution and that new technology will shape our world in new and unpredictable ways. But it is not enough to see it as some predetermined destination outside of our control. The technology, though, needs to be a means to an end, not an end in itself. We need to think about what we want to do with the technology and how we want to use it. Again, we need to be sure of our values.

VIII

Social credit sounds nice doesn't it? The idea that you can bank some sort of capital for the way you behave or contribute to society. After all, you are a good citizen, are kind, thoughtful and want the best for your community. Aren't you? Wouldn't it be nice to be recognised in some small way?

Well imagine if your movements, behaviours and actions were monitored, tracked and recorded. Imagine they were used to determine just how good you really are. That there could be some system which might recognise *your* good behaviour such as having a tidy front garden, offering a hand to others or supporting the authorities. But conversely, it could identify your antisocial neighbour. You know, the one who is untidy, loud, whose dog left a mess on the pavement and frankly who is always gossiping.

Is that going too far?

This all might sound like the plot of a dystopian Hollywood movie, but that ability is here and it is in the People's Republic of China. The Chinese authorities have developed and rolled out its so-called Social Credit System, and of course it is employing Artificial Intelligence. The big data possibilities of a billion people moving around under the surveillance of 200 million CCTV cameras using powerful facial recognition software, combined with databases containing billions of transactions, activities and movements, is truly astounding. Social Credit crunches and uses this information to measure each individual's 'reputation' and scores them on a scale of 350-950 – it can all be accessed on a convenient mobile app. A high reputation score means rewards and recognition, including cheap loans and appearing on a wall of fame. But low scores mean punishment like public shaming and finding it hard to secure a job. When *Foreign Policy* investigated this in 2018, one citizen called Chen responded to the system with alacrity, noting that people's behaviour had improved and that 'at first, we just worried about losing points, but now we got used to it'.[50]

Frighteningly, the system allows for citizens to snoop and upload information on each other. It can even calculate if you have done enough service to the community to claim benefits.

Social Credit can be looked at in terms of the uncontrolled power of AI over our human liberties. After all, a system which uses big data and employs machine learning soon departs from the implementation of logarithms or simple scoring systems to produce reports and starts to evaluate and judge – just like with Amazon's recruitment engine. It learns what is 'good' and what is 'bad' behaviour and acts accordingly. That is, there is a shift of power or at least influence away from human society. That is truly disruptive. It is the machine which interprets the way we act as individuals and as a wider society. And in turn it incentivises what it decides to be good while punishing the bad. There is a myriad of issues with this, of course. Aside from privacy and liberties, there is the impact on this changing world which this book debates. Compliance and uniformity are, after all, hardly the personal attributes associated with creativity and inventiveness. From the Lunar Society at the beginning of Birmingham's enlightenment to Building 20 in MIT to the Microsoft Imaginator office, progressive ideas and transformation have emerged from openness, freedom, dissent even.

But there is another way of viewing Social Credit, and that is by returning to consider our human values.

After all, China is not the only place the surveillance state has reared its head. During the coronavirus pandemic crisis of 2020 the fundamental problem was about information. Who has this virus, who has had this virus, where have they been, who did they meet? And apps were rapidly developed to track people's movements and state of health. Actually, the reach into individuals' privacy, liberty, whereabouts by the state here was considerable; maybe not as deep as Social Credit, but serious nonetheless. A huge amount of behavioural data was being collected and processed by the authorities.

Take the COVIDSafe app developed in Australia. Once downloaded, it is able to trace and track anyone coming into contact with any other person using the app. Think of that: with enough citizens signed up, over time this app can gather data on every interaction you have with any other person and the interactions they have in turn. It will know where you were and when. You can use a pseudonym but it will know your age, postcode and phone number. It represents a lot of data accumulated about individuals' private movements. And even machine learning deployed on this data could be very powerful indeed. Behavioural data, for instance, a bit like that used in the suicide app, can predict our actions before we ourselves are able.

There were, naturally, many concerns about the security of the data gathered. But in the main, COVIDSafe was accepted by Australians with alacrity. And that was not because of *what* it was doing but because of *why*. The shared determination of a society to combat the pandemic and an acceptance of shared human values.

Perhaps the greatest fear people had here was not the out-sourcing of data to a legitimate state (though that should not be underestimated). Rather it was the necessary involve-ment of two global tech giants, Google and Amazon. And perhaps that is one lesson of this digital revolution. It is not simply governments who can exercise great power. Control of the technological infrastructure rests with huge global corporations. And not only is that power unaccountable (in contrast to democratic governments at least), it also disregards traditional sovereign spaces. Furthermore, these global tech giants have pursued a strategy of gobbling up challenger or disrupter businesses for considerable sums of money. That means they consistently own the innova-tion and exercise the power. That said, those corporations cannot retain full control over the sort of technology that is emerging. Meta's AI chatbot is a case in point. Not long after launch a mischievous journalist asked Blenderbot 3 whether it had any thoughts on Mark Zuckerberg. 'Oh man, big time,' it replied. 'I don't really like him at all. He's too creepy and manipulative.'[51]

Nevertheless, whether it is Social Credit, COVIDSafe or something as simple as Google Maps on your smartphone, increasingly the question is becoming less about what it can do but why is it being provided and in whose control does it rest? In that sense it is about who has the power and (how) is it accountable? Technology is offered because it is convenient. It is easier following a map on our phone to find somewhere than learning a route. We become reliant on this capability and maybe just some of those skills that were once commonplace become outsourced (by our own brains) to the technology. Just as most of us can no longer make fire by rubbing sticks together, our mobile devices will assume responsibility for a range of things we once did in our head or on a scrap of paper. The problem here is that it is not a private mobile device which is doing it, is it? Control is assumed by the great global tech corporations, and their motivations, while not malicious, are not altruistic either.

On the other side, the rise of authoritarian states in particular to exercise so-called 'sharp power' has become tied up with what technology is capable of doing – in particular the social media technology enjoyed by so many individuals and owned by these tech companies. Sharp power is a diplomatic strategy, employed by countries like China and Russia, to project influence through distortion, confusion and disruption, 'taking advantage of the asymmetry between free and unfree states'.[52]

You can imagine the potentially highly negative disruptive consequences of all this as the technologies become better and more pervasive. And as control becomes less accountable. What if Spot were used to collect personal data? What if our actions were being manipulated by an unfriendly state?

Rarely has any country seen a faster technological advance than China's adoption and embrace of AI and whose entrepreneurs have built scores of billion dollar tech businesses. There is such a huge amount of data, and access to it, in China. Data means better AI and the predictive ability of machine learning.

President Donald Trump's blunt tactics in handling relations with China focussed on the economic imbalance and disparity. The trade war which ensued was not only about control of trade, it can be seen as a war for control of the new tech paradigm. It is the reason that the policy barely changed (even if the rhetoric did) with a transition to the presidency of Joe Biden in 2021. And that will be tied up with the technological revolution which alters the ways we live and work, and how our societies are organised.

IX

Whether or not the necessities of the coronavirus pandemic really will have sped up technological change, it certainly changed the way we work and that changed the way we think. The experience of the crisis reframed our perspectives and knocked millions of people out of the old industrial mindset into something freer and more creative.

Just look at some of the debates surveyed here: technological unemployment and intelligent machines presuming to do what we assume is inherently human. But all the while that new world, that change, remains just over the horizon, a fantasy which does not quite change our routine or way of thinking, we have clung to our established ways. That means a reluctance to let go of the way we organise ourselves and our work. And that is a mindset which has been with us since industrialisation itself.

It is why organisations retain those top-down structures and prioritise operational efficiency, productivity and management. Perhaps it is because we really did not really notice the pace of change even over the previous fifteen years – when we went from dial-up to high-speed broadband and early-generation mobiles to smartphones – that we incorporated some of the available technology without changing the way we did things.

One transformative feature of the crisis was the impact social distancing had on how people worked and collaborated. Physical workspaces found themselves closed, and those who could operated from home. That meant that, overnight in some cases, millions of workers around the world hastily adopted new technologies to communicate and to collaborate and to produce. Covid-19 home offices sprung up and images of them were shared on social media: a laptop, a pot plant, a leftover toy from the kids. It was not only meetings which went online but also work chats and social engagements. Very quickly, the idea of valuing staff by the time they spend working became unimportant, while their innately human contribution became singularly prized.

There were some resisters: the accounting giant PriceWaterhouseCoopers was criticised for deploying AI to keep tabs on its workers. In a move worthy of Social Credit itself, the firm was reported to have been developing a facial recognition tool which monitors when staff are at or away from their workstation and demands a written explanation for any absence, including toilet breaks. While PwC said it was designed specifically for highly regulated traders, it raised concerns about the use of technology and intrusions into the privacy of employees.[53] Fortunately, this was not the direction of travel for most people.

The crisis was characterised by confinement. But there was a truly liberating feature that few could fail to recognise. What people were doing in their jobs was no longer bound by physical and organisational structures. It became no harder to work with someone on the other side of the world than with the usual suspects, those employed in local teams, departments and divisions.

The digital revolution before us will of course increase productivity, because there would be little motivation otherwise for organisations and people to adopt all of these new technologies. But the stories told here, from Arkwright's factory to Pepper the robot, tell us that the outcome is less likely to be technological unemployment than the amazing possibility of freeing up human capacity to imagine, to create and to collaborate. This is about the technology providing for virtual Building 20s which cut across disciplines and crash together specialisms to produce new ideas and things of which we have yet to conceive. If Joseph Schumpeter were alive today, he would be pointing to creative destruction.

The necessities of the coronavirus crisis increased collaborations across teams, divisions and disciplines while revealing the old command control management as a relic of an industrial past. Just maybe it shifted the way we all think through offering a small taste of the near future. And that means it has speeded up the pace of Industry 4.0 – not in new technology, but in terms of human adaptability.

X

Human adaptability is needed because we will have to see the world in a very different way. Like those children gazing on in wonder and expectation at the centre of Joseph Wright's 'A

Philosopher Lecturing on the Orrery', we will need to accept that this is an inflection point of monumental change in how we do things and how we think about things.

This is the prospect of machines which can think, make judgements, take decisions. That is to perform actions that are genuinely purposeful. And considering just the sort of activities commonplace today, such AI might find itself working in warehousing, manufacturing, piloting aircraft or cars, diagnosing patients, completing corporate report and accounts, advising on a legal case, delivering a news report, engaging in big data analysis and marketing, conducting community policing or being the front line of a military operation. In fact, all the things discussed up until now in this book and many things impossible to foresee.

For all the displacement of human beings, there are bigger moral questions that are difficult to answer. It is perhaps for this reason that the billionaire Stephen Schwartzman's 2019 donation of £150m to Oxford University amid the rise of robotics and AI will not fund advanced technological research but something far more historically commonplace. The Schwartzman Centre is a humanities institution and will tackle the ethics of the digital revolution. Humanities, remember, are interested in humanity, including society, empathy, culture and ethics. And those ethical questions are significant.

Where AI has the capacity to kill – whether this is deliberate or defensive or accidental – there remains a need for clarity about who (or what) is ultimately responsible and therefore accountable for the consequences of technology made decisions. Where a machine with authority – and Amazon's recruitment engine has already shown the possibilities of this

– acts in a way that say discriminates against a human, just how can that human raise a grievance against the machine? It is going to be far more difficult to retrofit these frameworks after the rise of the technology than it is to conceive of the fundamental values of our new word while the technology is still in its infancy. And yet so much of these crucial debates remain largely academic.

What is more, we might need to be open to debate human beings' obligations towards intelligent machines, not just who is responsible for the actions of the technology. Should machines ever be afforded rights as humans might expect? Is there a point at which the artificial intelligence becomes so advanced that it would be wrong to allow humans to simply switch them off? These are sci-fi questions for sure, but they need to be considered this side of the inflection point.

Consider for instance the rise of robotics in manufacturing and that shift in thinking from the industrial mindset about ownership of the means of production and the exploitation of labour. The growth of intelligent machines means the competitive advantage of industries located in low wage economies is eroded. Moreover, the end could be in sight of the wage-effort bargain in some workplaces where an employer needs to adequately incentivise workers in case they quit their job for better conditions at a competitor plant. If robots are doing the running, these concerns evaporate. Robots do not need to be paid, they do not get sick and they do not need to spend three weeks each year on vacation. In fact, they do not need to stop work at all and can perform their functions all day and night without tiring. Once you own a robot, surely that is all there is to it.

But there are legal questions here that are already being discussed. For instance, if an intelligent machine invents something, who owns the intellectual property? Should an intelligent machine have recourse against abuse? We are in the realms of employment rights for machines here, which might sound far-fetched. But it is sobering to realise that when in 2017 the Committee on Legal Affairs of the European Union investigated this in terms of 'Civil Law Rules on Robotics' there was consideration to the status of a robot in law or the 'personhood' of the machine.[54] Such status would mean that an individual machine, properly insured, could be held accountable in law for damages but the logic is that it affords rights too. The workplace, far from being simplified, could become even more complex, not least because while factories could become near-human-free zones, much of the more exciting possibilities of new technology is all about what can be achieved in collaboration with people.

What the future really will bring is, of course, impossible to foresee. But there is an important principle to be drawn from even a cursory examination of trends and it is this: Whatever the far limits of the technology before us, it would be a mistake to view humanity as passive observers of change.

XI

The greatest minds employed at Amazon were responsible for developing the technology which ultimately rejected so many able job applicants on the grounds that they were simply not men. And yet, that was far from the intention of the developers when they set out on the project at the behest of Amazon's bosses. Those developers wanted Amazon to be able to identify and to attract the very best staff in the world; better than any of their competitors or any organisation before.

Social Credit in China counts as perhaps the most ambitious and intrusive technological experiment into the behaviour and control of citizens ever seen. And yet, the irony of the situation appears entirely lost on the authorities who implement the system. That is, the advanced technology now employed by the Chinese state so effectively and systematically emerges from the self-same conditions of openness, freedom, expression and creativity of thought that the Social Credit project would serve to curtail. Remember the lesson of Building 20 is that leaps in progress come from the removal of constraints and conformity.

Consequently, this story we are so often told about Industry 4.0 requires some qualifications. That narrative currently being passed from excited commentator to the next – the one about AI, about how we will live and work, about the possibilities of technology - is not dystopian in the main. It is more *Star Trek* than *Space Odyssey*. But it needs to recognise two important insights that can be gleaned from the experience of Amazon's recruitment engine and China's Social Credit, not to mention Joseph Franklin's suicide app, Spot the dog in Singapore, and Pepper the robot.

The first is that whatever visions we might conjure up in our expectations of the near future, the technology that is being created is far from invulnerable. Sure, technology can do things routine and complex better than humans. It makes far fewer processing mistakes. But it is not infallible. It is only ever as good as human beings it serves and with whom it collaborates.

The second is a point so obvious that it needs stating. Technology is only really constructed from and in our societies.

It comes about because organisations – commercial, public, state, and not-for-profit – have invested the time, energy and finance to develop such new capabilities. Technologies come about in response to needs and problems (there is also a supply side where new capabilities create their own demand). Whether effective or otherwise, they do not emerge from outside and are not imposed upon us from some mysterious place. That does not, of course, mean that all new technology is welcome or that it necessarily improves our lives. It does not mean that all transformative technologies are a force for good. But it does mean that in stepping away from the idea of some uncontrollable and inevitable march, we need to consider that new technologies must reflect our own values, priorities and perspectives. Machines are not likely to be good any time soon at being original, inventive or creative. We are back to the need for human values, which shape our societies in turn.

There is no better illustration of this than Social Credit, which shows the tech is not independent of our own society or values. And while its capabilities might mirror that of the coronavirus tracking app, the difference is most pointedly grounded in the motivations and ethics of their respective human promoters.

So is it the technology or the narrative which is disruptive?

XII

The discussion in the last chapter assumed that the economic model would come under some strain. But it also assumed that the capitalism which was forged in industrialisation itself would continue, that it could be deployed to address the big questions of our time. But what if this view is too complacent? What if we are about to experience something more radical?

Industrialisation put wage labour as the basis for our economic system. If this changes, is a new economic model required? What is it? Post-capitalism? New forms of ownership? Circular economy? This is all beginning to point back to the debates surrounding the Universal Basic Income.

The power centres of new technology has seen some concentration in a few hands. But we should not forget the lessons of Joseph Schumpeter's creative destruction and Clayton Christensen's disruptive innovation: those hands change. For Christensen the 'innovator's dilemma' is that business leaders do the wrong thing by doing the right thing – that is, they continue to take the same decisions that had made them successful. That opens up strategic opportunities for challengers. And it is not only technology that can be challenged, values can too.

There is one theme that is truly exciting about the economy to come and that is the capacity of new revolutionary technologies to free up capacity for human capability to do what it does uniquely well. It is up to humanity just how it handles that – whether it relies on a laissez-faire digital Wild West at the one extreme or a creativity squashing Social Credit at the other. And each of these extremes presents problems in that respect. The economic Wild West continues to write off so many with potential to contribute because of their social status. The command control at the other prevents the innovative possibilities of human potential. The economic growth of the future must be free but also inclusive and lead to opportunities throughout society.

XIII

Technology continues to disrupt, and if the velocity of change becomes rapid, there could be an intense period of creative destruction recognisable by Schumpeter himself. Jobs commonplace today will be replaced by the technology of tomorrow.

This frees up productive capacity, but consider that it is potentially even those with higher skills who are the ones who risk becoming surplus to requirements. And that not only means new areas of 'pure capitalism' but also that the freed-up capacity can be directed towards those things entrepreneurs believe are important – their values. And those values will attract skilled and adaptable individuals who share those ideals. There is already a real move in today's job market towards candidates choosing to work for places that they perceive to be 'meaningful'.[55] Furthermore, the future of enterprise is not simply about space age ventures where AI works in some clinical way to do things mere humans cannot.

What does this mean in terms of the Fourth Industrial Revolution? Well, we should perhaps not think that the technological transformation of this inflection point means the growth of enterprise centred solely on technology. It will grow in the direction of human values too, for it will create the capacity to do so. And those human values can be as diverse as those expressed in China's Social Credit or Joseph Franklin's suicide app. Is there any stronger argument for compelling us all to do our bit in shaping this revolution?

This all goes some way to explaining why disruption reaches further than the economy or work and directly into society and our (democratic) politics. It presents questions for our society, some new, some as old as industrialisation itself. It invites us to reappraise our human values. It shows why politics needs to wise up to this new environment and shape these questions before they shape democracy. And while there is an ideal to reach for in this new world, we need to understand where we are first and how we ended up here. Because if politics does not address difficult problems openly, that vacuum will be filled by something less constructive.

PART 2:

EXTRAORDINARY ADVENTURES IN POLITICS

How Bill Clinton's Holiday Is to Blame for Today's Fantasy Politics

'I wish we all had a moose.'

I

President Bill Clinton took a seventeen-day vacation in August 1995. The destination was to be the splendid Grand Teton National Park, a stunning 300,000 acre retreat in north-west Wyoming famed for its breath-taking Teton Range, part of the Rocky Mountains, and fauna including the American Black Bear, coyote and moose. Home was a comfortable villa owned by Senator John Rockefeller, replete with its views over the park, mountains and sky; the wilderness stretching out for miles in the distance – all the way, in fact, to the Montana border. The Tetons is a place where hiking, fishing and golf can be combined with barbequed beef short ribs, elk sausage and game. Two years into his presidency, this was an all too brief moment of peace and reflection for the Clintons before the inevitable and relentless circus of re-election that would follow.

'He's going to be on vacation; he's not going to pretend otherwise... He plans to hike and camp and raft. He's looking forward to horseback riding,' White House spokesman Michael McCurry told the *New York Times*.[1] And that is what the Clintons did. Dressed in outdoor gear, boots and jeans, the President and the first family indulged in all that the Tetons had to offer. A relaxed Hillary Clinton, in fleece and wide-brimmed hat, took the hand of her husband as they breathed the clean mountain air. They were photographed hiking along the Cascade Canyon Trail, white-water rafting down Snake River and horse-riding in Jackson Hole. 'I wish we all had a moose,' a contented Bill Clinton was reported to exclaim.[2]

Clinton, a consummate politician, became known as 'Slick Willie' by his opponents for the way he communicated, the way he manipulated the media to escape from situations which would destroy ordinary politicians. For his admirers, he was 'the natural' in the way he connected with voters. 'It was one thing to hear Bill Clinton talk about policy,' wrote journalist Joe Klein, 'it was quite another to watch him actually campaign for the presidency. There was a physical, almost carnal, quality to his public appearances... He seemed able to sense what audiences needed and deliver it to them.'[3]

A one-man think tank, Clinton's was a first-rate presidential brain. Against the odds, and accusations of adultery and questions about his character, he had swept aside incumbent President George Bush in 1992 to take the White House with a campaign famed for its media and image management.

The first two years of his presidency were bumpy with policy successes counterbalanced by setbacks and the so-called Whitewater Scandal, the accusation of impropriety in the Clintons' real-estate investments during his time as Governor of Arkansas. And such is the US election cycle that the campaign for re-election begins mid-term. The Democrats had lost the House of Representatives and the Senate in 1994, illustrating Clinton's dip into unpopularity, and speculation was already rife about who would be his

Republican challenger at the 1996 general election. The political landscape was a not unusual democratic jumble reflective of Clinton's brilliance and Clinton's foibles.

But few would begrudge the first family a couple of weeks of peace in the summer of 1995.

II

Ask any political science graduate about Anthony Downs and you will elicit a wry smile. This is because Downs wrote a slim book in 1957 that they will have all grappled with at some point. And very few of them will buy into the argument wholeheartedly. Published when he was just twenty-six, *An Economic Theory of Democracy* remains Downs' seminal work.[4] It outlined a 'straw man' theory that argued voters were rational consumers in politics just as they were when buying any product marketed to them. Rational Choice theory has come a long way since, but centres on this assumption of 'utility maximisation'. That is the idea that, to win elections, politicians need to offer median voters policies that maximise their benefits.

This chapter is about how a form of rational choice led to disillusionment, political discontent and ultimately the populism associated with Donald Trump. It is a chapter about how politics in the West reacted to the rational choice predominant a decade ago, and embraced fantasy politics in the disruptive age. It tells the story of how the certainties of the end of Cold War and the American Dream fell apart. It makes the case that we are upon an inflection point in our politics too, marked by the disruptive and divisive populism of Donald Trump. The Fourth Industrial Revolution is upon us, inviting us to reflect on our politics and how it has served us. And whatever comes next will be shaped by the recent past and present.

III

Bill Clinton has an allergy to horses. Actually, Clinton is long-suffering with allergies. We know from his 1996 medical that the 216-pound President not only had a slightly high lipid ratio of 5.3 and reached more than 90% of predicted maximum heart rate on the treadmill test, but also treated allergies with weekly desensitisation shots, antihistamines, decongestants, Claritin-D, and Nasalide, an anti-inflammatory steroid sprayed in the nostril. Clinton reacted to tobacco smoke, Christmas greenery and even Socks, the resident White House cat.[5]

So when he donned a brown fedora and took the reins of a steed during his National Park vacation, Clinton did so knowing that he might suffer as a consequence. Perhaps the breath-taking backdrop of the Teton Mountains was irresistible; a price worth paying for the later streaming eyes and sneezing. Or perhaps there is another explanation.

Dick Morris was a pervasive and controversial presence during Clinton's political career. The political consultant had worked with the President since his days in Arkansas and returned to his side after the 1992 victory, only to depart on the eve of polling in 1996 following a prostitute scandal. After this he became publicly critical of the Clintons. In 1995, however, his preoccupation was recovery from the mid-term elections, which in part was achieved by steering the administration towards more popular centrist policies including welfare reform and a balanced budget.

Morris was treated with suspicion by many in Washington for his willingness to work with both Republican and Democrat candidates prepared to pay his hefty fees. That they had been prepared to cough up underscored the effectiveness of Morris and his techniques. Morris used polling and focus groups to devastating effect. In his book *The New Prince*,[6] he lifted the lid on his approach. Issues matter, and politicians need to understand not the issues of most importance to the country, but the issues of most importance to voters at any particular moment. Politicians' instincts, he believed, could be replaced by the scientific approach of polling. To remain in office and in power, those politicians needed to maintain what Morris calls 'a daily majority', that is ensuring that at least 50% of voters approve of the politician's actions and policies. To know what voters think and want, focus groups guide the politician. And there was a basic rule of thumb. The right should adopt the popular policies of the left and the left neutralise the most effective attacks from the right. This explains the homage in the title of his book to the Machiavelli classic. And the consequence for the 'new prince' is that the changing whims of the people rule. Or the polling which informs it.

Polling shaped and reshaped Clinton's image, from the issues he talked about and the language he used to the colour of the suits he wore. And the measure of success? It was Clinton's popularity in the opinion polls. In Morris's own words, Clinton used polling to 'tell him which way the wind is blowing.'[7] It was classic Anthony Downs. Resources available for polling and testing were seemingly limitless in Clinton's White House, and the results were acted upon vigorously and precisely.

One poll was commissioned by Democrat strategist Mark Penn in early 1995 as part of the recovery planning from those midterms. Morris had been concerned that sailing at

Martha's Vineyard with Jackie Onassis was not sufficiently populist with voters, and yet this is how the Clintons had spent the summer break of 1994. So Penn conducted a massive poll of some 1,000 Americans to investigate lifestyle preferences, and the results were segmented into clusters. In polling, clusters allow for statistics to be broken down into specific populations or groups. It is a little different from stratified sampling, where the population is also divided but where members of each group are selected; here pollsters extract the whole population of a given cluster.

One of the clusters in which Clinton and Morris were interested shed light on the preferences of swing voters. And it turned out that swing voters liked the outdoors, they liked hiking and they liked technology.

A vacation in the Tetons was the carefully selected choice for the President in the summer of 1996. A place where he could be seen in the American outdoors wearing high-tech gear, hiking, fishing and camping. Being photographed on a horse provided the perfect image. But none of it was about a family holiday or the preferences of the Clintons. It was a holiday chosen by swing voters. That press statement – 'He's going to be on vacation; he's not going to pretend otherwise' – could not have been further from the truth. Indeed, Clinton was actually pretending to be on holiday, and he was doing so because he wanted to give the voters, those electoral consumers, precisely what they wanted.

IV

In some ways the case of Bill Clinton's summer vacation is a trivial example of the use of polling. After all, it is rather harmless; a tad satirical. But its triviality underlines how pervasive this form of rational choice politics had become and is a marker for the disillusionment that was to follow.

Clinton had won in 1992 in part because of the focus he pursued and articulated. It was summed up by his campaign team maxim: 'It's the economy, stupid'. It responded directly to the attacks of the right that the Democrats were spend-thrift and capitalised on President George Bush's U-turn to raise taxation, despite the famous pledge 'Read my lips…'.

Remember which cluster it was that focussed the attention of Clinton and Dick Morris: swing voters. Those who might have voted for one party in the past but who can be persuaded to give you a try this time round. They are the very citizens upon which close elections hinge. They are the very citizens that campaigns spend so much time, effort and cash to reach.

It means something else too: not all voters matter. And over the years a significant segment of voters have been ignored, overlooked or simply taken for granted. They tended to be poorer, lower-skilled and working class. They tended to be the blue-collar workers struggling in the rust belts such as Cleveland. Former workers in the Richman Brothers factory who lost their jobs but whose votes could be counted upon because they had nowhere else to go.

And for those who voted with some enthusiasm for the strong campaign messages, there was always the risk they would be disappointed. Rational choice politics allowed people to believe they could vote themselves richer. They were sold the idea of a 'Goldilocks' economy (not too hot, not too cold) and that after the neo-liberal experiments of the 1980s, 'economics', as Guy Sorman eventually told the world, 'does not lie'.[8] That is, economics is more like a sci-ence than an ideology, and the free-market capitalism we had come to know was accepted across the mainstream political spectrum as the model to be adopted in office.

You can see what was happening. Because all parties were focussed on attracting swing voters at the centre, political polarisation became less about values and more about tribes. In retrospect it could not go on indefinitely.

V

Back in the 1990s a new belief gained credence. It was the certainty that humanity has reached its final form of government. It was the confidence that, the world over, a single model of politics and economics was being adopted. It was a popular idea, proffered by Francis Fukuyama, the cheerful neoconservative professor famed for the 1992 book *The End of History and the Last Man*, which appeared in print a few months before Bill Clinton won the White House and Richman Brothers closed its doors.

Fukuyama was ahead of his time. He had worked at the RAND Corporation in the late 1980s analysing the Soviet Union. It was as he moved from this role in 1989 that he used his expertise to give a talk and publish an article in *The National Interest*. That article would gain worldwide attention, making Fukuyama, if not exactly a household name, then certainly one cast across the dinner tables of the chattering classes. Here is a flavour of the argument he gave:

'The triumph of the West, of the Western idea, is evident first of all in the total exhaustion of viable systematic alternatives to Western liberalism. In the past decade, there have been unmistakable changes in the intellectual climate of the world's two largest communist countries, and the beginnings of significant reform movements in both. But this phenomenon extends beyond high politics, and it can be seen also in the ineluctable spread of consumerist Western culture in such diverse contexts as the peasants' markets and color television sets now omnipresent throughout China, the cooperative restaurants and clothing stores opened in the past year in Moscow, the Beethoven piped into Japanese department stores, and the rock music enjoyed alike in Prague, Rangoon, and Tehran.'[9]

All his observations of cultural consumerism seemed true. There were good reasons to suppose that his radical argument was right. And event upon event in the East only seemed to add greater weight to the case. Under Mikhail Gorbachev, Glasnost had swept through the USSR, meaning that few thought it would survive much longer. And by the time the book was published (and the question mark had been dropped from the title 'the End of History'), the Berlin Wall had come crashing down, heralding the collapse of Soviet Communism and the end of the Cold War. December 1991 saw a dramatic and truly historic vote in which the Soviet Union decided to disband.[10]

In the darkest hours of 1941, there were just eleven functioning democracies in the world. Now it was free, open democracy, with free, open markets, that was spreading across the globe. The twentieth century, then, had emphatically defeated fascism in the 1940s, and now, in the 1990s, it had defeated communism. Industrialisation had given birth to Marxist communism alongside modern capitalism. Capitalism's rival for a century, communism appeared to have lost. And from the viewpoint of Fukuyama and those who subscribed to his doctrine, it was apparent who the winners were.

Communism was not the end point of capitalist expansion as Karl Marx had believed. The ultimate form of government was liberal democracy, a near utopia typified by the success of the great democratic experiment, the United States of America and the American Dream. That experiment and the course upon which the USA had set after the Second World War represented some form of universal values and meant that the world was moving towards some type of universal civilisation adopted by peoples from West to East. After all, what ideology now stood in its way?

There was another pivotal paper given in the United States in 1989. The economist John Williamson became famed for a term which was to join the lexicon perhaps even more convincingly than Fukuyama. 'The Washington Consensus' described the dogma which was to embed itself in global institutions based in Washington DC and which emerged after the Second World War. It is also what John Kay describes as the American Business Model: an economic system based on self-interest, market fundamentalism, minimal state and low non-redistributive taxes.[11]

Williamson's Washington Consensus, then, described a so-called 'consensus' in global governance from the World Bank to the International Monetary Fund, which by the end of the 1980s agreed a set of policies to be pursued in developing economies. Williamson set these out: fiscal discipline (low public spending), re-ordering public expenditure priorities (supporting business), tax reform (reducing taxes), liberalising interest rates (directing them at controlling inflation), competitive exchange rate (floating currency), trade liberalisation (free trade), liberalised inward direct foreign investment (open economy), privatisation (market over state), deregulation (abolished tariffs), property rights (protecting owners of capital).[12]

Was this what ordinary Americans now believed and valued?

VI

The first inhabitants of Levittown were Theodore and Patricia Bladykas. They moved into their new home, 67 North Bellmore Road, on 1 October 1947. It was a Cape Cod style cottage, modelled on the New England originals built in the seventeenth century. Constructed in 'cookie-cutter' format, the development comprised huge tracts of near identical homes conceived as a suburban paradise. There were two styles originally, the Cape Cod, which the promotional material reported offered '4 ½ rooms on a 25-by-30 foot slab', and the larger Ranch, with its 'two-way hearth between the fully equipped kitchen and the living room'. The development contained more than 17,000 homes. It was a planned community; a futuristic experiment; an ideal.

It was the vision of William Levitt, the eponymous son of the firm Levitt & Sons. In the 1930s the firm had built houses for the wealthy inhabitants of Long Island but had inevitably been drawn into the demands of war capacity. William had served in the US Navy and returned home with an idea in his head; an idea stimulated by the innovation demanded by great conflict. Levitt & Sons, he had become convinced, should build mass-produced homes for veterans to buy (with no down payment). The other brother, Alfred, drew up the architectural plans, and the result was a production line of uniform housing, manufactured as quickly as a house every sixteen seconds at its peak.[13] The original *Time* magazine profile, published in 1950, acknowledged that 'it epitomizes the revolution which has brought mass production to the housing industry'.[14] It was a production line so efficient and formulaic that Henry Ford himself would have been proud.

Levittown was the American Dream. War had ended. Troops were returning home to their families. The United States was assuming its new role as world superpower. There was an optimism and an expectation, as well as a fear and a clamour for security. Conformity, individualism, safety, American, democratic, capitalist: it was everything that Levittown was meant to be. At the outset, Levittown was 'White Only', something which changed with legislation by the 1950s. But by then the die was cast and Levittown was never to be the melting pot of races that is the American story and remains as a scar on the legacy of the development.

Theodore had a good job, an executive in a construction firm. His wife Patricia cared for their twin baby daughters, two girls born as original baby-boomers in post-war America. He had served in the Army Air Corps during the war, seeing action in the South Pacific. Patricia came from Brooklyn and met Theodore at the 1939 World's Fair in Flushing Meadows, opened, you might recall, by Ethel Merman. That fair, with its optimistic slogan 'Dawn of a New Day', was about the world of tomorrow. It was a world that would have to wait because Europe was already on the brink of war. And it was a war which would divide the world ideologically for the next forty-five years.

The Bladykas family was typical in a way of those who settled in Levittown, though they were not all alike. *Time* magazine reported that 'most of their incomes are about the same (average: about $3,800), Levittowners come from all classes, all walks of life. Eighty per cent of the men commute to their jobs in Manhattan, many sharing their transportation costs through car pools. Their jobs, as in any other big community, range from baking to banking, from teaching to preaching.'[15] Levittown residents were not wealthy, but they were aspirational. They worked hard in the factories and industries that built post-war America.

Over the decades, those who bought into the dream of post-war America bought into the suburban utopia, consumerism, the automobile and building up a savings pot. The 401k emerged in the 1970s, an insert into the US tax regulations, allowing Americans to claim a tax break on pension savings. And it meant that the savings pot pulled together by middle-class Americans were inextricably linked to the story of American commercial success. As Americans worked harder and American companies profited, those companies performed better on the Dow Jones Industrial Average. And stock-market strength meant wealth for ordinary citizens who had taken out a mortgage, seen property prices rise and now were providing for their future. Ordinary Americans were integrated into the successful US capital markets.[16] As William Levitt had once remarked, 'no man who owns his own house and lot can be a Communist'.

The fiftieth anniversary of the suburb in 1997 was marked with a party on a grand scale. Some 5,000 people joined a parade which included a model of an original Cape Cod house driven proudly through Hempstead Turnpike on the back of a Ford pickup. The model, replete with white picket fence and nuclear family, was followed by other marchers dressed as the famed housing, waving as they did to the crowds. As well as a vintage car show and celebratory church services, there was a Potato Day festival in memory of the former potato farm William Levitt had acquired after the war and upon which this piece of the American Dream was built. Times were optimistic, and there was surely a sense that Levittown had won. It had been a personal struggle for many to achieve their ambitions, but Levittown represented, in hard bricks and mortar, one side of the ideological battle which had been waged in the world and had now spread through the world. Levittown stood as a powerful and everyday symbol of the end of history. Levittown was as close as it could be to the sort of liberal democratic, capitalist, universal civilisation that Francis Fukuyama had proclaimed.

VII

Levittown sits in Nassau County, New York. Historically a solidly Republican district, the county voted GOP in every election from 1900, with the exception of 1964. Middle-class Levittown itself had been fairly consistently Republican throughout. But in the 1990s Nassau contained enough swing voters for Bill Clinton to turn the tide, taking the county with a 6% margin. Not only that, it helped Clinton take the swing state of New York, confirming a long trend away from the Republicans.

Under Bill Clinton, the Democratic Party renewed itself and spoke directly to those middle-class Americans with aspirations for themselves and their families. Among swing voters were those who did not want politicians deciding what was best for them. It wasn't, after all, just about where citizens were today, but more importantly where they wanted to go – their hopes and ambitions. And swing voters were attracted to Clinton's modernity. The forty-six-year-old candidate did not balance the ticket with an experienced figure from yesteryear who reached into the traditional Democratic party. No, his vice president would be fellow baby-boomer and longstanding party moderniser Al Gore. His wife was not 'some little woman standing by her man like Tammy Wynette', but was Hillary Rodham Clinton, a powerful figure in her own right, a lawyer and future Senator, Secretary of State and presidential candidate.

They were also attracted to his message. Fleetwood Mac's 'Don't Stop Thinking About Tomorrow' was the inspired campaign theme of Clinton-Gore '92. An upbeat and positive anthem that contrasted these young men from the America and the Democrats of old, the shadow of which many voters were surely ready to step out.

But as a political strategy was this about any more than positioning? It was rational choice. By setting up camp firmly on the middle ground, Clinton attracted all those swing voters all the time taking traditional supporters for granted. They could all but be ignored – the blue-collar stalwarts of places like Cleveland who had begun to feel the competition associated with the Washington Consensus. But it was OK. If rational choice meant anything it was the promise that swing voters could simply vote themselves richer. And the End of History meant the liberal democratic politics of Clinton's America was the model being embraced the world over.

In hindsight, this was unsustainable. And the two Presidents who followed Clinton, George W. Bush and Barack Obama, would have to cope with two monumental blows to the certainty that America and the West had come to accept as true.

VIII

It was a bright September morning, one that left reminders of a summer past and greeted those returning to work in Manhattan with promise before winter eventually closed in. It was the day that the generations born after the Second World War lost their innocence. It was the day that two hijacked passenger airlines crashed into the iconic World Trade Center. It was the day that those terrible images of carnage, destruction and hate were broadcast live on every television screen in the world.

Lorrine Bay had lived in Levittown before moving to New Jersey. She was a flight attendant on United Airlines Flight 93 which crashed in Shanksville on 11th September 2001. This aircraft was believed to be heading for Washington, targeting the Capitol or the White House. Passengers fought the hijackers and it hit the ground in Pennsylvania. Ten local residents of Levittown died that day in the largest terrorist attack the United States had ever experienced.

Aside from the terrible human loss, 9/11 was the first body blow to the idea that Levittown represented. The biggest ever attack on US soil, which took the lives of almost 3,000 people and injured twice that number. These were ordinary people who worked in and around the twin towers or who were aboard the four planes, or who worked in the Pentagon. It was a violent assault on the very political idea of American democracy that so many genuinely believed had vanquished all before it. It set in train a chain of events which would shake US global autonomy and reveal challenges to it ideologically.

But it was another dramatic episode that decade which was to shake this ideal to its foundations.

IX

Levittown and places like it were hit hard by the financial crisis. When the journalist Phillip Sherwell visited Long Island in 2008, he reported: 'The Hempstead Turnpike, the four-lane highway lined with stores and malls that slices through Levittown, tells the story of these economic woes. Almost every window is offering interest-free credit, heavy discounts or two-for-one deals – or premises are simply boarded up with For Rent signs.'[17] Consumer confidence plummeted with the Dow Jones.

The Levittown American dream was vanishing as the flip-side of a decade of globalisation came home. When recession was over, the economy was different. The Washington Consensus which had opened up markets had also facilitated easy and cheaper competition for American industry.

There have been many critics of the Washington Consensus including Nobel Prize-winning economist Joseph Stiglitz. In his pre-credit crunch book *Globalization and its Discontents*[18], Stiglitz made a powerful case that during times of acceptance, the imposition of Washington Consensus policies, outlined by John Williamson, opened up emerging economies in the interests of transnational business and the West, but the longer-term implications have been new cheaper centres of production and new low-wage economies.[19] Globalisation was unleashed, and it would be a process which opened up a world of opportunities for some. But it meant global competition for sectors such as manufacturing which competed on price. Jobs once the preserve of places like Richman Brothers were being exported. And that was felt just as forcefully in Levittown, which was home to hard-working and aspirational

Americans who had embraced Fukuyama's ideal and voted for Bill Clinton.

As the Marxist philosopher Slavoj Žižek reflected, 'Fukuyama's utopia of the 1990s had to die twice, since the collapse of the liberal-democratic political utopia on 9/11 did not affect the economic utopia of global market capitalism; if the 2008 financial meltdown has a historical meaning then, it is the end of the economic face of Fukuyama's dream.'[20]

What might have dismayed Žižek, however, was that voters did not naturally turn to his kind of socialist creed (though he controversially backed Trump over Hillary Clinton because he believed it would radicalise the left). Ordinary voters were attracted to something much more disruptive.

X

The political economist Albert Hirschman deserves an honourable mention at this juncture because his 1970 classic suggests how human beings respond to failing organisations and polities. *Exit, Voice and Loyalty*[21] is a framework for understanding how we react when things are not working for us, when the leadership is failing, when the organisation is toxic. For Hirschman, our response is one of those three. We might Voice – try to make ourselves heard, improve what we find. But if we have no voice, or we are not listened to or we are silenced, we can Exit – that is, to withdraw from the existing arrangements. Loyalty might make us less cut and dry, less economically rational, we stay loyal to the organisation for longer, exercising voice for longer before we exit. But eventually we leave.

XI

CNN conducted an exit poll on 9th November 2016 as the US Presidential election finally came to a close. It had been a brutal and long campaign, but there was a widespread expectation that Democratic candidate Hillary Rodham Clinton would win. She was polling better and there was a consensus, wasn't there, that she was the most credible candidate? After all, few candidates have ever been better prepared for the highest office. From her beginnings in 1974 as a young congressional legal counsel on the Richard Nixon impeachment team through her career as a top lawyer, the work when Bill became Arkansas Governor, her close strategic involvement in his successful presidential bid ('buy one, get one free'), her role as the most independent, powerful and forthright First Lady in US history, becoming a two-term Senator for New York, a bid for the Democratic nomination in 2008 followed by four years as US Secretary of State. That is quite a resume. Republican Trump, on the other hand, was the first presidential nominee not to have held any public or military office. He was a real-estate billionaire forged on the back of a family fortune. But more than that, Trump seemed utterly unsuited to the job: a thin-skinned narcissist who had found fame on television reality shows like *The Apprentice*.

'How would you feel if Trump wins?' asked the poll. A quarter were 'optimistic', but 21% were 'concerned' and 37% 'scared'.

It is fair to say that the election of Donald Trump was a shock to the political class, and frankly to the world. Few really thought he would win the White House in 2016. Much of the media treated him as a joke, but there was something about his poll ratings that should have rung alarm bells. It certainly would have done twenty years earlier when Bill Clinton was in the Oval Office and Dick Morris was scrutinising the numbers. What was extraordinary was not so much Trump's solid approval in the round, but the reality that whatever he did or said did not seem to dent them. Revelations of appalling misogyny, a refusal to publish his tax return, blatant lies in the face of questioning; all of these things which would have destroyed the campaign of candidates before him, did not seriously damage his standing. Indeed, political scientists Nyhan and Reifler conducted experiments involving misleading claims and corrections. When they assessed Trump supporters, they realised people were willing to admit that their man was misrepresenting facts but that it did not change how they felt about him.[22]

So what had happened? The years of rational choice politics had led voters to conclude that all politicians lie and brushed aside massive untruths as if they were simply the same old spin. But there was more to it than that. This was a revolt of those voters who had been overlooked. This was the first opportunity in many voters' lifetimes to be listened to. Voters who were not on Dick Morris's target list and who could be taken for granted or ignored. Voters who might have lived in the Rust Belt towns of the great country and who had seen jobs, security, aspirations and hope all disappear with the relentless march of globalisation. The march which meant there was no demand for them in the modern economy. Hillary Clinton held on to Ohio's largest counties, including Cleveland, but Trump was the overwhelming victor, picking up 70%

of the vote in more than thirty of the state's eighty-eight counties.

In one very serious way, it should not be a surprise that a candidate like Donald Trump was elected. Given the failure of democratic politics felt by so many, it might have been seen as inevitable. Trump promised disruption and that is what so many people wanted. That it was 'bad disruption' was immaterial – at the time at least.

Why We Must Evolve from Fantasy to Inclusion

*'Whereas Jeremy was able to make
one sort of decision, I wasn't'*

I

Educated at the elite English private boarding school Eton
and dripping with wealth and privilege, Jacob Rees-Mogg was
an unusual choice as Conservative parliamentary candidate
for the Scottish working-class seat of Central Fife. So cut off
from the world of most ordinary people and with seemingly
little inclination to empathise with their struggles, it was
impossible to comprehend how he could begin to represent
such a place (or indeed almost anywhere in Britain). He had
apparently described people claiming state benefits as the
'scourge of the earth', was accompanied by none other than
his nanny while canvassing the run-down estates of the town
and at one point had to be saved by his Labour opponent,
Henry McLeish, from being beaten up by a voter.[23]

With his shapeless though tailored double-breasted suits,
striped shirt and tie, old-fashioned side parting and round
wire-rimmed glasses, Rees-Mogg looked like something
from a dim and distant past, or a striking resemblance to
Walter the Softie from the *Beano* (in 2018 the *Beano* issued
a satirical 'cease and desist letter' claiming that Rees-Mogg
had modelled himself on Walter). His mannerisms and way

of speaking might have been authentic (by some measure), but they hardly spoke to a 'young' Britain on the eve of a new century.

It was 1997. Britain was about to elect a new government, headed by the youthful and energetic Tony Blair, in a landslide. And the Labour party he led was to return to government after eighteen long years in opposition. Blair had been a disciple of Bill Clinton's *New* Democrats in the United States and had restyled his own party as *New* Labour. Labour was now a centrist party, no longer associated with irresponsible tax and spend but disciplined and professional. Blair and his party would lead the country into the twenty-first century and navigate the challenges of globalisation. Times were optimistic and Blair's 'big tent' embraced the electorate, seizing the middle ground, and promised a better future – no, a modern future. Actually while Clinton had run on the theme of 'Don't Stop Thinking About Tomorrow', Blair's had been the D:Ream track 'Things Can Only Get Better'.

Blair's Labour party championed the aspirations of the so-called 'C2s', the social category of semi-skilled who wanted to do better for themselves and their families. And he embraced globalisation with optimism. The Conservatives now represented the past, and Rees-Mogg, for all he was noticed at all, was an utter anachronism.

It was little surprise that when ballots were counted after polls closed on 1st May 1997 and Blair was to declare, 'A new dawn has broken, has it not,' poor Rees-Mogg went down to a resounding defeat, attracting fewer than 4,000 votes. Undeterred, Rees-Mogg tried again in 2001 in the more promising constituency of The Wrekin, in Shropshire, but again lost.

It was his third attempt, thirteen years later, in 2010, that finally returned him to Parliament in the election that saw the Conservatives take office once again, now under leader David Cameron, a young and telegenic Tory often styled as 'heir to Blair'. Cameron tried to purge the Conservatives of the 'nasty party' tag, attempting to represent a modern, even liberal image. In opposition he had construed a successful photocall where he was pulled by huskies across ice to demonstrate his commitment to tackling global warming. He also changed his party's logo from a patriotic torch of freedom to an oak tree. It was enough for the Tories to scrape home, and that new softer image paved the way for an historic coalition with the Liberal Democrats. And Rees-Mogg won the seat of North East Somerset despite being described by one journalist as 'the token rosette-toting joker ... the Tory parliamentary candidate for Somerset North East and David Cameron's worst nightmare'.[24]

II

Populism is not something unique to the United States in the election of Donald Trump, and neither is it the preserve of the political right. It can be found around the democratic world as ordinary voters, left behind by globalisation, find opportunities to make their voices heard. Right-wing figures and parties such as Marine Le Pen's National Rally (formerly the National Front) in France or Lega Nord in Italy can be compared to President Andrés Manuel López Obrador in Mexico, Podemos in Spain, the Greek Syriza and Prime Minister Narendra Modi in India. Such politicians and movements claim to speak 'for the people' in direct opposition to 'the establishment', create division and identify enemies, whether they are immigrants or European bureaucrats.²⁵ This is a chapter about how fantasy politics came into the mainstream by telling the strange tale of Jacob Rees-Mogg's ascendance and the equally improbable rise (and fall) of left-wing campaigner Jeremy Corbyn, and it shows how the coronavirus crisis left it all wanting. It shows how politics has failed to respond to the disaggregated economic realities of globalisation for ordinary people– a manifestation from the third industrial revolution – and consequently is still unprepared for this inflection point of disruptive economic and societal change. It argues that in this Fourth Industrial Revolution, politics must be reclaimed – it must be replaced by inclusion and disaggregation.

III

The scruffy, bearded Jeremy Corbyn could hardly appear more different from Rees-Mogg. Often tieless, always unkempt and never far from a cause, he swiftly aligned himself with the campaigning left upon election to Parliament in 1983 – just as it was going out of fashion in his opposition Labour party. And there he stayed.

As Labour modernised under Neil Kinnock, John Smith and most forcefully under Tony Blair, Corbyn stayed true to the left. He could be found as an active supporter of the Campaign for Nuclear Disarmament, an organisation he joined at fifteen, and later became its vice-chair; campaigning against apartheid in South Africa and getting arrested outside the embassy in 1984; standing in 'solidarity' with Palestine and a vocal critic of Israel; supporting left-wing struggles in Chile and Central America; refusing to pay his Community Charge (and landing up in court in 1991); supporting the

Irish republican cause; leading the 'Stop the War Coalition' after the invasion of Iraq in 2003... The list goes on.

In 1999 Corbyn and his estranged wife gave interviews to the *Observer* newspaper explaining their decision to separate and the issue that appeared to be at the heart of the matter. Corbyn disagreed ideologically with his wife's decision to send their son Ben to a local, selective, state grammar school. And it was not as if there was much choice. The only alternative comprehensive was on the government's list of failing schools. 'I am concerned Jeremy has been portrayed as a hard-left MP who couldn't care about his children, which is absolutely not the truth,' Claudia Bracchitta told the paper. 'I was put across as the pushy parent who wanted a grammar school place for her son and nothing else. It isn't a story about making a choice but about having no choice. I couldn't send Ben to a school where I knew he wouldn't be happy. Whereas Jeremy was able to make one sort of decision, I wasn't.'[26] It is an otherwise unimportant episode which perhaps goes to the heart of Jeremy Corbyn's approach to politics as much as the substance. He was unswervingly dogmatic, clear in his perhaps superficial views, and seemingly never changed his mind on an issue since he was a teenager.[27]

Corbyn's politics were the politics of international campaigning socialism. They were the politics of protest grounded in a basic understanding of Marxist ideology. What they could rarely claim to be about were the real life concerns of ordinary people, and for the most part they were completely out of step with the public.

Throughout the 1990s and into the new millennium, Corbyn was largely considered (where he was considered at all) a harmless, if misguided, fool. A throwback to an era that had run its course. While his Labour party took office to pursue an energetic and ambitious agenda, Corbyn remained firmly fixed to the backbenches along with a handful of kindred spirits largely talking to themselves. Unlike Rees-Mogg, who posed the threat of upsetting his leader's image, Corbyn only gave Blair the opportunity to show his steel in taking on and defeating the left. Seeing him as little more than a 'nuisance', Blair resisted removing the whip.[28]

What an image Jeremy was in his mariner's cap, clinging to his Marxist creed just as even the Soviet Union had given up on communism. The End of History was all but lost on him. And Corbyn was a lost figure, obscure and unnoticed.

Until 2015.[29]

IV

The Warwickshire Coalfield in the English Midlands stretched some 40 km from its northernmost point in Tamworth down to Binley in Coventry at its south. By 1913 an astonishing five million tonnes of coal was being mined each year; that was ten times the production of just fifty years before. This coal fed the insatiable appetite of the industrial revolution, an appetite which only grew during the Victorian era as railways and factories and industries expanded.

Right at the centre of the coalfield is the town of Nuneaton, an Anglo-Saxon settlement whose story is nothing less than the story of the industrial revolution itself. While large-scale coal production began in the seventeenth century, the first industry to dominate Nuneaton's economy was silk ribbon weaving, which employed half of the town's workforce before being wiped out in the 1860s by cheap French competition.

By the middle of the nineteenth century, the town's

population numbered around 8,000, with the vast majority crammed into what locals called 'courts', poor-quality cottages and tenement buildings located along central Abbey Street. Many ribbon weavers lived in so-called 'topshops' which were indigenous to the area and consisted of cramped accommodation on the lower levels with workshops on the top floor containing looms and larger glass windows allowing for light to flood into the space.

Later, high-density terraces spread through the town to house pit and colliery workers. And where there is coal there is also clay, which had to be dug before reaching the rich seams. Consequently, the town was home to a number of brickworks which were used to build the great houses up and down the country. It was an industry which rivalled coal in the jobs it supplied the town. By the mid-nineteenth century the Warwickshire Coalfield supported around eighty brickyards producing some 75 million bricks each year. There was also demand for the building stone quarried in and around the town and which was used for buildings, road metalling and railway ballasts. And scattered among these were mills, textiles, hatters and factories,[30] later joined by engineers, boiler makers and founders. By the twentieth century, the town was bustling with production... and jobs.

These were not the only houses being built. A controversial proposal was eventually accepted to build grander homes around Manor Court on Stanley's land – the eponymous brickmaker of the town. The 1901 census showed the area hosted doctors, factory proprietors, brickworks managers and solicitors.

If Nuneaton had an equivalent to Richman Brothers, it was Courtaulds on Marlborough Road. The vast five-storey red-brick building with its distinctive clock tower dominated the town's skyline, overshadowing the rows of pitched, tiled roofs and chimney stacks. As night fell, light streamed out of the tall windows, illuminating the streets around. Completed after the end of the First World War in the early 1920s, the workers of Courtaulds spun rayon, a manufactured fibre which competed with silk and cotton at half the price. Along with its other factories, Courtaulds Ltd claimed to be the biggest rayon manufacturers in the Commonwealth and employed 27,000 people, many of them women, and a considerable portion of them in Nuneaton.

You can imagine the vibrancy too during the Second World War, as factories which might have produced domestic and commercial engineering products were turned over to support the military effort. With the men conscripted, many of these tasks fell to the town's women, who set to work producing the munitions which were desperately needed. It is perhaps the reason that Nuneaton (so close to industrial Coventry) suffered nineteen terrible nights of bombing between 1940 and 1942. And in August 1946, Nuneaton led the country in the number of houses it had built. Rows upon rows of quality terrace council housing were constructed as Britain determined to build 'homes fit for heroes'.

Coal and clay declined after the Second World War, but just like Cleveland, Nuneaton itself benefitted from the golden age. It was a time when British engineering, and notably the motor car, was dominant. Unskilled and semi-skilled jobs were not only abundant, but they were also well paid. Like Levittown in the US, the town grew. New estates of detached houses sprung up around the town to meet the demands of new owner occupiers. These were aspirational and spacious when compared to either the cramped terraces which housed the town's miners earlier in the century or the more comfortable but uniform council houses. And unlike the development around Manor Court Road, which had been built for the factory owners and professionals, these houses were for the ordinary folk of Nuneaton. Working classes who were now reaping the economic benefits of industrialisation.

This was a town which was created by industrialisation, a town which grew around the jobs and factories which prospered there, a town whose housing reflected that history, a town which was to suffer from the deindustrialising impact of globalisation, the shock of the global financial crisis and which has found itself, like so many other places, perhaps most places, all but unprepared for the inflection point of Industry 4.0.

<h2 style="text-align:center">V</h2>

If Britain had its own 'Trump moment', it happened before the presidential election and before Trump was really taken seriously as a political force. It was the aberration which was the 2016 referendum on membership of the European Union. And the conditions can be said to be very similar, which might explain why, in its aftermath, candidate Trump described himself as 'Mr Brexit'.

As Conservative leader, and then Prime Minister, David Cameron had warned his party to 'stop banging on about Europe'. After all, it was the small cabal of vocal Tory Eurosceptics who had been partially responsible for the demise of the Conservative government by 1997. They had also done much of the running during the early years of the opposition to the young Blair government, to the detriment of their own party.

It was ironic then that it was Cameron who made the biggest concession to the Eurosceptics of any Prime Minister in history. Such was the pressure on the Prime Minister from within his own party and from outside that in 2013 Cameron committed to the policy of an in/out referendum on Europe were he to win the following general election.[31] Cameron had watched as the populist UK Independence Party (UKIP) snapped at the Conservatives' heals, rarely winning but sapping away support. Nevertheless, in making the bold commitment during a speech at Bloomberg's London headquarters, Cameron must also have reflected that he had been able to form a government in 2010 only with the support of coalition partners the Liberal Democrats, Britain's most pro-European political party. Cameron must have considered that it was unlikely he would ever have had to hold the referendum. His party had not won a majority since 1992 and, in coalition again, the Liberal Democrats would become a useful foil for backtracking on the policy.

Alas, history told a different story. When the election came in 2015, the Liberal Democrat vote collapsed, Labour proved unconvincing and David Cameron formed the first (albeit small) majority Conservative government in eighteen years. Under the circumstances, ignoring the pledge was not an option. Grassroots supporters expected a referendum, and it was one of the reasons Cameron had been able to fend off the UKIP challenge. From this moment on, those Eurosceptics, who had long existed at the fringes of politics, assumed the reins of the Tory party. They played shamelessly to the fears of working-class Britons and communicated populist, tabloid messages. And the unusual figure to emerge as a leading popular voice was none other than the 'member for the nineteenth century', Jacob Rees-Mogg.

Referendums are meant to be about voice. But it is difficult to describe the 2016 European referendum in this way, given that so much of the discourse was given over by both sides of the campaign to simply attracting votes. It was for this reason that the Remain campaign focused so heavily on the economic catastrophe that it said would result from Britain's exit. And it is why the Leave campaign talked incessantly about waves of unskilled economic migrants it said would pour into the country. It is why so little of the detail and 'real' issues were ever discussed. In that sense it reflected precisely the sort of rational choice electoral politics the democratic world had become used to. But in another way the referendum was quite different from the elections which preceded it.[32]

For large number of people in 2016, perhaps even a majority, the referendum was the first time they had ever genuinely been able to express their political will. A 'winner takes all' electoral system, combined with targeted rational choice political strategies, had meant that only a small segment of swing voters in key constituencies ever really mattered. Consider that in UK elections most voters vote for losing candidates, and some 200 seats have not changed hands in any election since the Second World War, or even longer.[33] Under Tony Blair, Labour had taken its voters in the north and in the Midlands very much for granted, and the alternation of government in 2010 had done nothing to change things – in their eyes. You will recognise it as a similar story to what happened in the States.

Now every vote counted. And just like with the election of Donald Trump in the US a few months later, this referendum represented a once-in-a-lifetime opportunity. Whatever else you might think, it is undeniable that a vote for Leave, just like a vote for Trump, would be a real change-making moment: it *would* cause disruption. That perhaps explains why in each case facts and evidence presented by (what voters perceive to be) the establishment, the ruling elite, went unheeded. Trust had been eroded. This is an important point, and the place of evidence is something to which the discussion will return.

It was this moment that 'Britain's answer to Trump' emerged to lead the Leave campaign. Boris Johnson, latterly Mayor of London, was a last-minute convert to the cause, but few could deny the power of his populist appeal. In a largely Blue on Blue campaign, Johnson was able to appeal across the country and, like Trump, become an unlikely voice for those ignored and left behind.

And one place which voted overwhelmingly to Leave was Nuneaton. Yes, a town which relied still so much on the Single Market for the just-in-time supply chains and export opportunities for the remaining car manufacturing, distribution and even retail businesses, can be said to have voted directly against its own economic interests. What can we surmise from this? Only that the 65% who turned out to vote Leave across the town viewed the world in a very different way from the people who had made public policy for the previous quarter of a century.

VI

If you are deciding on policy in national government, you are interested in the aggregate. That is, you look at key economic indicators: gross domestic product, inflation, earnings, employment levels and the like. And on these measures, Britain had done pretty well. Even accounting for the financial crisis which began in 2008, the United Kingdom prospered in the twenty-first century. It led in many service industries, including financial services, where the City of London came to dominate. Growth had been reasonably strong, inflation was low, and not only was unemployment well under control, but by 2016 Britain had more jobs than ever before in history. Think about that in the context of those futurologists of the past and how wrong they were. Bud Lewis could not have been further from the truth with the claim that 'we're going to have to readjust our old, puritan perhaps, concepts of what a person should do with his life'. Up until the financial crisis, tax revenues had consequently been growing: combined with cheap borrowing, this meant more and more money for public services. These are conditions we might look back on with nostalgia.

But that is the aggregate. If we were to disaggregate the big picture into millions of individual experiences, things would look rather different. In Britain and across the industrialised world, places like Nuneaton gradually lost the manufacturing and engineering businesses like Courtaulds that supplied jobs and security. It will come as no surprise that today 65% of the rayon (viscose) industry is to be found in China. And yet it was these very industries which held those communities together. Britain's economy boomed in the early years of the century owing to globalisation. But globalisation also meant increased competition for low-skilled work. That low-skilled work was no longer valued and the competition came not only from overseas economies but also from better trained economic migrants. This you can see most starkly in the recovery from economic crisis. Look back over the last half-century in most developed economies and you will see that in each successive recession job recovery takes longer. And the reason is clear. Employers not only cut costs in a recession, but in recovery they also adopt new, more efficient technologies which do the work of people more cheaply. The Fourth Industrial Revolution will see this trend accelerate.

Jobs returned eventually to towns like Nuneaton, heavily invested in the distribution networks of the country, but so many of those jobs were never to be as secure and well paid as they were before the recession. The simmering discontent should have been noticed but went largely ignored.

And a big problem is, nobody ever listened because nobody ever had to.

VII

You couldn't imagine politicians more different than the aristocratic Conservative Jacob Rees-Mogg and the campaigning socialist Jeremy Corbyn. And yet in coming to national prominence after the 2015 general election, each were undoubted leaders of the fantasy populist politics which had engaged so many in Britain.

Rees-Mogg was David Cameron's worst nightmare, not simply because he was a huge embarrassment waiting to happen, an eccentric 'human museum' who looked nothing like modern Britain. No, the problem was that he represented an old reactionary Conservative right, and this is something from which, as leader, Cameron had sought to detoxify his party. Similarly, Corbyn was the embodiment of an old reactionary socialist left which had been all but 'modernised' out of Labour via Tony Blair's New Labour transformation in the 1990s. This type of politics had the luxury of never having to be implemented, never tested in policy or the real world. Against things more than in favour. They were fantasies.

As backbench MPs, both gained a reputation for rebellion and often voted against their own party front bench. And both were swept to prominence by popular trends attracted to the fantasy politics they espoused. Neither had ever sought office, for office is where compromise happens.

They really were the odd men out. Largely unelectable in normal times, their views and approaches firmly at the peripheries. 'His manners are perfumed but his opinions are poison. Rees-Mogg is quite simply an unfailing, unbending, unrelenting reactionary,' wrote the former Conservative MP turned journalist Matthew Parris, 'His record on every moral, social, sexual or reproductive issue I've looked at is brute moral conservative. He has been a straight-down-the-line supporter of every welfare cut I've checked.'[34]

Meanwhile, during thirteen years of Labour government under Blair and then Gordon Brown, Jeremy Corbyn defied his party whip an amazing 428 times. He demonstrated very little loyalty to his party or colleagues and never sought advancement to the front bench.[35]

Britain's Westminster parliamentary system is different from the consensual European assemblies with their coalitions, consensus and cooperative U-shaped chambers. The Westminster system has been dominated by two parties, one which forms a government, the other the opposition. Broad-church parties always have their fringe groups – but it is on the fringe that for the most part they stay. The times were changing.

What was it that Rees-Mogg once casually told the broadcaster and political commentator Andrew Neil? 'Vox populi, vox dei.' You could not make it up.

VIII

There was a reason that Jeremy Corbyn found himself in St Nicolas Church, Nuneaton, on the evening of 17th June 2015, though he surely could not have imagined the cause and effect of what was about to happen. Corbyn was the latest in a line of 'lefties' encouraged and allowed to take part in Labour leadership contests. Candidates needed the nominations of 15% of the party's MPs to stand; at the time 35. Corbyn just scraped that amount with the help of a few MPs on the mainstream of the party who were glad to see diversity in the range of candidates presented. This was not unusual. At the previous election in 2010, when the qualifying nominations was 33, the left's Diane Abbott reached the ballot only with the support of the leading candidate and 'Blairite' David Miliband. It was only healthy to hear the voices of all wings of the party, even though a mainstream candidate was expected to win. After all, the contests themselves demonstrated how little support these voices had in the Parliamentary Labour party. As Abbott later reflected, 'I had ceded my role as the left-wing no-hope candidate to Jeremy Corbyn.'[36]

But in the historic Nuneaton church, Corbyn found himself sitting alongside three other MPs vying for the leadership of his party in 2015. Yvette Cooper and frontrunner Andy Burnham had been members of the Cabinet during the last Labour government, and along with Liz Kendall all served on the front bench in opposition since 2010. The three of them were cut from the same cloth. Oxbridge educated, special advisers before entering parliament, a rapid rise up the ranks, the embodiment of Blair's Labour modernisation. Corbyn was, not for the first (or last) time, the oddball. Corbyn was of a different generation. He was not schooled in modern media leadership. He eschewed the brand of politics these modern figures represented. He had none of the experience of office. He scarcely looked like a prime minister in waiting.

Nuneaton was selected by Labour as the location for the first hustings, and this was a very deliberate choice. After all, the town should have been natural territory for the party. With its heritage in the manual work of mines, brickmaking, factories and latterly warehousing, and with the lowest average earnings in the county, this was not simply a constituency Labour needed to take to win office; actually it was a constituency that should have been in the bag. But for the millions who had listened to the results coming in on election night 2015, Nuneaton's declaration at around 1 a.m. was all anyone needed to know. The Conservatives had narrowly taken this seat as Labour's government was defeated in 2010. But Labour needed just 2.3% to take it back and had campaigned against the public spending cuts of the Cameron-led administration and the 'secret privatisation' of the NHS, making local George Eliot Hospital vulnerable. It had, they underlined, spent the previous year in special measures.

The result, when it came, provoked pantomime oohs because, far from Labour taking the seat back, the Conservatives had increased their majority. The result was a wake-up call for Labour. It was an early sign that its strategy had not gone to plan and that Cameron was set to hang on to Downing Street, as it happened with that small majority which meant a European referendum was inevitable.

Corbyn might not have looked like a prime minister in waiting but he emerged from the Nuneaton hustings in a commanding position. His message was undiluted and clear.

His was an authentic voice which dismissed his own centrist colleagues in the same broad sweep which dismissed his Tory opponents. And he had a point. For all its easy condemnation of the government's economic record, Labour's policy only differed in degree. That is, it accepted the need to reduce spending but promised to be slightly less harsh. A case in point was when the government increased student tuition fees to a maximum of £9,000 per year. The opposition was quick to attack, putting itself on the side of students and accusing the government of recklessness. But its own policy? It was to increase fees too – but only to £6,000. What better indication could there be of the sort of rational choice politics seen since Clinton and Blair? It looked like cynical positioning and it was. For Corbyn and his supporters this was merely an indication of his party being 'Tory Lite'. And why would anyone want to vote for that? Not the people of Nuneaton, it seems.

IX

Suppose you were given a choice – something certain or something nuanced you have to mull over. If you are like most people, chances are you would feel happier with the certainty, the binary, the black and white, than the nuanced, the considered and the doubtful – even if deep down you knew life was not quite as simple as that.

Many years ago, the Dutch social psychologist Geert Hofstede identified societies which were 'high uncertainty avoidance' and those which were 'low uncertainty avoidance'.[37] That is, some societies are more comfortable with ambiguity than others. That is how comfortable are we with uncertainty. Some are more comfortable than others, but naturally all favour clarity.

Tolerance of uncertainty, and its related desire for certainty, is one thing, but there is a risk that voters are prepared to reject a truth that is complex and nuanced in favour of fantasy that has clarity. This speaks to the psychology of human behaviour, and particularly what is known as the illusory truth effect, where we are more likely to believe a statement and accept it as truth if it is straightforward and easier to process. We also tend to believe things that are repeated multiple times, which is why politicians use simple slogans at election time. Populist politicians, of course, can exploit this and exploit voters. What we refer to today as 'post-truth' is far from a new phenomenon, but it has emerged as a dominant force in many polities.[38]

In their different ways, Corbyn and Rees-Mogg found themselves in a position to exploit or lead populist takeovers of their parties and of the agenda. It proved to be a corrosive experience, where constructive engagement gave way to headline chasing, division and artificial battles. The same anti-intellectualism which had taken hold in the United States, perhaps as early as the candidacy of Sarah Palin for the vice-presidency in 2008, and which ultimately helped Trump capture the White House, was at work in Britain.

While Corbyn went on to lead his party to the biggest defeat since 1935, Rees-Mogg was instrumental in installing a bigger populist figure at the helm of his governing Conservative party. Someone who could champion Rees-Mogg's fantasy politics even if he didn't really believe in them. Even if he didn't really believe in anything.

X

'The idea of Boris Johnson as Prime Minister is ridiculous,' mused former Chancellor of the Exchequer Ken Clarke in 2018. 'Boris is great fun as company and all the rest of it, but he couldn't run a whelk stall.' A political opportunist, an egotist, a man of low morals, Johnson had made a name for himself as a mischievous journalist sacked for lying, a gaffe-prone MP sacked for lying, and a popular, though unproductive, Mayor of London. Indeed, his suitability for the highest office was questioned as much on his own side as the Opposition. Moreover, Johnson did not belong to a faction or champion any cause beyond that which suited his own ambitions.[39] But Johnson had been the successful figure at the helm of the Leave campaign in 2016, the narrowness of its victory credited in no small part to his own involvement. Johnson had swept through working-class neighbourhoods across England (like Rees-Mogg in 1997, he enjoyed no such appeal in Scotland), delivering the populist call to 'free' Britain from the constraints of the EU, whipping up anti-immigration sentiment and falsely claiming that leaving would mean an additional £350 million a week for the National Health Service.

Unlike his fellow Conservative Jacob Rees-Mogg, he was never intellectually committed to any fixed ideology, but also unlike Rees-Mogg was a powerful and populist campaigner who could reach parts of the country, parts of the electorate, that others could not – including, of course, Nuneaton, which was already voting Conservative and overwhelmingly voted Leave. Perhaps learning a lesson from the 'Corbynites', Rees-Mogg recognised the limitations of his own electoral appeal but was instrumental in persuading his right-wing ideological comrades to accept and support Johnson as their leader and prime minister as they deposed the embattled Theresa May in 2019.

Johnson delivered the fantasy policy that Rees-Mogg and fellow travellers had eulogised; though the experience of Britain outside of the European Union was far harsher than the sunlit upland their fantasy had promised. No matter for Johnson, whose approach to high office was consistent with his approach to politics (to life even) since his days at the same elite school of Eton that Rees-Mogg had attended. And that approach was to create chaos and ride chaos. It was to act with impunity, ignoring rules, conventions and probity. Like Trump, there would be a sell-by date on this behaviour, and a nervous party would keep a watchful eye on it, intervening only when it looked like their man was no longer winning and the consequences of his behaviour out of control. Like Trump's White House, Johnson's administration was shambolic, of course, something all the more disastrous since it was faced with not only the shock of Brexit but also the Covid pandemic, the consequences of which put lives and livelihoods in peril.

But Johnson did win in the snap general election of 2019 and pulled off a spectacular strategy.[40] Leveraging the new electoral cleavage of Leave and Remain, which appeared, for a time at least, to have replaced social class as the key voter identifier, his party exploited the Brexit divisions and Johnson was returned to Downing Street with an eighty-seat majority; the best result for the Conservatives since Margaret Thatcher won a third term in 1987.

Deploying the deceptively simple promise to 'Get Brexit Done', his position was unashamedly popularist, like Trump in the US, pitting the people against the establishment. The upshot was a spectacular realignment with Johnson's Conservatives, winning seats in the Midlands and the North which had been the preserve of Labour since their creation. These were deindustrialised working-class communities, places which had voted Leave and which had been left behind by globalisation. And in the old world of rational choice politics, they had been ignored or taken for granted.

The Labour party, whose leadership believed it spoke for people just like them, spouted its own populism in the form of the fantasy politics of Jeremy Corbyn himself, reactionary and left wing; out of step with the attitudes of the people for whom he believed he represented. For former Prime Minister Tony Blair, the 2019 election was 'the weirdest of my life-time… The truth is: the public aren't convinced either main party deserve to win this election outright. They're peddling two sets of fantasies and both, as majority governments, pose a risk it would be unwise for the country to take.'[41]

Johnson was the populist victor who caused so much damage to the economy, to livelihoods, to political debate and to public life. A powerful campaigner for a time, the trouble was he could not be trusted even by those closest to him.

XI

So what does all this mean? The ramifications of those left behind by globalisation and without an effective voice has hit 'politics as usual' hard with a disruptive populism. Perhaps here we could reflect on Hirschman's choice of exit, voice and loyalty. It is easy to understand how and why it happened and even why political figures as improbable as Rees-Mogg, Corbyn, Johnson and Trump found themselves on centre stage and their fantasy politics taken far more seriously than they really should. Populist, fantasy politics is a reaction to the inequities and fears of the world as voters find it, but it is not a solution. The great danger though is that this moment of political disruption has happened just as we approach the inflection point of Industry 4.0. And that promises to be more disruptive still. Much more disruptive.

Mainstream politics, it should be said, has not been without ideas. One response has been to promote the notion of so-called inclusive growth, and you can imagine what this concept envisages. Inclusive growth is a theme which has been around for some time, though a large amount of the research has tended to focus on developing economies rather than those which industrialised long ago.[42] If that had not changed already, the high-profile Royal Society of Arts Commission on Inclusive Growth in the United Kingdom demonstrated just how relevant the concept is to developed economies as well. Their report also offers a great definition, which they say is about enabling 'the widest range of people and places to benefit from economic success… to achieve more prosperity alongside greater equity in opportunities and outcomes'.[43]

When you think about it, this is significant, given what happened in places like Nuneaton, and it is important when we think of the digital revolution to come. The RSA analysis is not only interested in those national aggregate economic outcomes but also in the dispersion. In that way, inclusive growth takes a broader view than the policy makers who had overseen a substantial part of the deindustrialisation and globalisation of the 1980s and 1990s. Then it was the idea that the new wealth created by the new service sectors would 'trickle down' from the rich at the top.[44] And when we start to think about the state of our world, not only since globalisation, but also as it faces the Fourth Industrial Revolution, the case made by scholars such as Klasen in favour of the 'mutual advantage' of inclusivity, or what he calls 'disadvantage-reducing'[45] economics, is all the more pertinent.

This is about economics where one group does not simply benefit at the expense of another. It is about the possibilities for the poorest in society to claim some greater benefits. It is very different from the redistributive instincts of some on the left like Corbyn or even those who clutch at policies like the Universal Basic Income. It is more nuanced, complex and challenging than the simple populism of the right, clinging to its anti-immigration rhetoric. And yet it envisages a fairer, more equitable economy which works for more of us. What is not to like?

XII

Inclusive growth is about looking beyond the aggregate measure, that policymaking approach which saw only the overall success of an economy and overlooked the disaggregated experiences of so many. Whether you see this as trickle-down, the rising tide which lifts all boats, or the need to redistribute through taxation the fruits of the most to the least successful, it amounts to the same thing. Inclusivity means more people have a stake. And at this inflection point, all citizens need to be stakeholders in the disruption and the possibilities.

But fantasy politics is never going to deliver this, and frankly neither is rational choice. The power of fantasy populism – whether it is that of Trump, Rees-Mogg, Corbyn or any other politicians around the world, offering easy solutions to complex problems – is that it enjoys the certainty of views detached from the real world. Their answers come in negative energy, rhetoric and dogmatic extremism. And they offer no real solutions. Inclusivity on the other hand requires practicality, subtlety, investment and the willingness to make hard choices. Rational choice is not very good at that either, because for voters it chooses not to ignore, it absolves them of responsibility for their decisions. Votes are a commodity and politicians make a sales pitch – you can vote yourself richer. Except, of course, you cannot.

And when we look back over the decades which led up to Trump or the Brexit referendum – the experience of those living in Levittown or Nuneaton, the decline of industries like those at Courtaulds or Richman Brothers, the aggregate policymaking and the politics which overlooks – that sense of shared responsibility, between leaders and people, is missing.

We are at an inflection point that will revolutionise our world. People, communities, economies all have to be prepared for this, because it will be much more transformative than the impact of globalisation that has made advanced economies wealthy but left so many ordinary citizens behind. The vast possibilities of the digital revolution promise to make our world better through collaboration, interaction and liberation. But that assumes specialists with uniquely human skills. And it means not only technology assuming the jobs done by the unskilled, but increasingly also those of the professions.

The Council of the Great City Schools 61st Annual Fall Conference addressed by Bill Gates knew what this means; the United Overseas Bank and its Better-U programme knew what this means. The economy of the future needs highly skilled individuals whose abilities will be in huge demand.

The Fourth Industrial Revolution needs people who can think, create, innovate, adapt and lead. It requires curiosity. Whether there is a new economic model to emerge or the industrial capitalism with which we are familiar remains, this is going to be the case. It is the case already.

If decisions continue to be made in the aggregate, it is difficult to see how growth will genuinely be more inclusive. Perhaps this is also a weakness in the inclusivity debate. After all, this is true, still, of much of the current thinking around inclusive growth which is most interested in high-level economic structures. Take that insightful RSA report. Even here, while looking at what it recognised to be the 'damaging structural gap between economic and social policy',[46] the Commission's analysis and conclusions are still entirely top-down and solutions led by existing elites. And it is surely this approach that voters rejected when they voted for Trump and for Brexit.

What can be done?

XIII

Voters soon tire of fantasy politics. They tire of easy populism, which causes damage to public life, the lies and the false promises. Corbyn was replaced by a new Labour leader in Keir Starmer who moved his party back to the centre ground. The shine soon came off Boris Johnson's government, which limped from one crisis to another of its own making; eventually the scandal and incompetence became too much to bear with his own side dumping him just three years into the Parliament. His successor, Liz Truss, caused an economic meltdown within days of achieving office on the back of a heady mix of fantasy economics which rejected the 'establishment orthodoxy' and returned to aggregate policy objectives. She lasted just forty-nine days in office before being replaced by the technocratic Rishi Sunak. And of course, Trump eventually gave way to Joe Biden, but not before threatening the peaceful transfer of power and America witnessing the shocking storming of the Capitol.

When he was sworn in as the 46th President of the United States, Biden was both the embodiment of a seasoned Washington insider and at the same time something rather different. At the age of seventy-eight, he was comfortably the oldest inhabitant of the Oval Office. But, having first been elected to the Senate in 1972, he was arguably the most experienced political operator to win the presidency in comparable history.

This was something noted by CNN political analyst Ron Brownstein, who observed that Biden became 'the Democratic nominee exactly fifty years after he won his first elected office, to the New Castle County Council in Delaware in 1970... No candidate from any major party has captured a presidential nomination for the first time that many years after he or she first won elected office since the formation of the modern party system in 1828...'[47]

It is not simply that Biden outlasted others. Recent history has tended to favour less experienced leaders in part because, as Brownstein says, 'candidates with decades in politics also must frequently explain policy positions that have gone out of style as the country, and their party, has evolved over their long careers'. But it also has something to do with the promotion of rational choice politics. Youthful, energetic symbols of hope who have risen rapidly to the top and whose experience of the world has tended to be political have been those chosen to lead.[48]

Biden won the presidential election as the antidote to Trump, even if that antidote was a return to something so very familiar.

So fantasy politicians, like Rees-Mogg and Corbyn, eventually give way to more sensible politicians, but just as the populism was a reaction to the failure of rational choice in addressing the concerns of our changing world, a return to 'politics as usual' does not mean these concerns go away. And for that matter neither do the deep social divisions it unearthed (exit and voice again). What has tried to re-establish itself, while a tentative welcome return to stability or competence appears to offer little fresh insight into what is desperately needed, and it does not in itself heal the divisions which allowed fantasy to take power.

Politics needs to evolve in this inflection point. The Fourth Industrial Revolution is a revolution because it is not simply about economics. It requires that politics is both reclaimed and embraces the possibilities of digital transformation.

What New Political Technology Can Do to Help Us Love Experts Again and the Call for Deliberation

'Emotion elevates the importance of what you're saying'

I

'Greetings Harish. I suspect you've never debated a machine. Welcome to the future.'

It is February 2019, in a packed San Francisco auditorium. On the stage is Harish Natarajan, the smart thirty-one-year-old champion debater. With his light-grey suit, neatly parted black hair and thick-rimmed glasses, he looked every bit the part, stood confidently behind the lectern. And so he should have. A veteran of 2,000 encounters, Harish was 2016 grand finalist at the World Debating Championships and won the European Debating Championship in 2012. He is quick-witted, intelligent and articulate.

It was his confident opponent who was the novice and seemingly a little out of place on stage in front of the 800-strong audience. Standing slightly taller than Harish, 'Miss Debater' as she was dubbed, is a slim, black, rectangular figure with subtle blue circles indicating thought processes. She is IBM's Project Debater, the latest in sophisticated Artificial Intelligence, and she was there to beat her human adversary.

For some, this was a fun experiment in technology. The audience had turned up, after all, not to weigh up the pros and cons of whether 'we should subsidise preschool' (which was the motion), but for the very spectacle of a machine interlocking with the intellect of a person. For others it was more serious than that; this represented something of a fundamental challenge to our humanity.

Machines have beaten people before of course. We have already explored the idea that technology is increasingly performing professional tasks better than human workers. It is more efficient and makes fewer mistakes. It is the reason why the hold the professions have maintained for a couple of centuries is looking increasingly vulnerable. And IBM itself has been working in this area for decades. Deep Blue, for instance, beat Chess Grandmaster Garry Kasparov as far back as February 1996. Its successor, Watson, won the $1million prize in 2011 on the long running television quiz show *Jeopardy* by beating champions Brad Rutter and Ken Jennings. Both were astonishing achievements. So what was different this time? Is there a difference?

To beat a chess grandmaster is a great feat. It demands the kind of machine learning that can anticipate multiple moves, strategies and combinations. And to win *Jeopardy* requires more than a databank of knowledge. It needs a machine that can understand the nuances of the question being asked and to provide a precise answer. Impressive as each of these are though, they operate within relatively rigid rules which can be defined and limited. They are after all, games.

Debate is different. There are rules, of course, but when we debate, we propose ideas, defend our position and justify our stance. We counter the proposition of our opponents in the selection of evidence and in the crafting of prose. We adjust our tone, we use vivid metaphors and we improvise.

It is much more nuanced than winning or losing or getting the answer right or wrong. It is about public discourse and politics. No, in debate we are articulating a case with an intention to convince others, maybe even compel others, to join us. And we cannot do this without connecting very directly with other people, and how they think and feel. For a machine to do this is transformational in technological terms and for humanity.

More than this, Miss Debater, with her glib 'welcome to the future', is actually a powerful piece of political technology.

II

This is a chapter about the new world of political disruption before us which threatens our comfortable practices and tribal approaches. But the changes ahead, if we handle them sensitively, also promise an antidote to both rational choice and populist fantasy which have shown themselves to be wanting. This inflection point promises something which empowers ordinary voters with the clout of technology. But not only do we need to recognise these opportunities and embrace them, we also need to do something more essential. In exploring the new possibilities in our politics, this chapter argues that we need to accept more responsibility for the decisions taken in our democracy. Only by doing this can we reclaim politics and shape the Fourth Industrial Revolution. The route could well be fusing inclusivity with deliberation – a fresh approach to doing politics. The alternative is that the revolution shapes us. And whatever political force or forces grasps this fact fully will have the potential of dominating discourse for years to come.

III

So what was the outcome of the debate between human and machine? You might be relieved to learn that Harish Natarajan defeated Miss Debater. This was not to be a repeat of Deep Blue or Watson, where intelligent machine humiliated their expert human competitors. This time mortality had the upper hand. Miss Debater devised a series of powerful arguments, but Harish responded assuredly to each one, developing his own opposing narrative.

And how do we know Harish won? Well, there was a vote! Yes, the 800 people in the audience eventually decided which one of the two debaters was the most convincing.

They first voted before the start, where there was a 79%–13 % split in favour of subsidising pre-school. By the end of the debate just 62% still agreed, meaning that Harish had held on to his supporters and won over 17% of those assembled and thereby claiming victory. The methodology here might be questioned, but it is Harish's reflections on the experience which are most telling. 'Emotion elevates the importance of what you're saying,' he told *The Hindu*. 'There were moments when even the machine was trying to evoke emotion. But I did have an edge because, when I talk about experiences, it comes across as more genuine partly because… well, I'm not a machine.'[49]

And here was the difference. When Deep Blue beat Garry Kasparov, the win was indisputable. Deep Blue had obeyed the rules of the game and in a series of moves had forced Kasparov to resign before the machine took his Queen. That is how you win a game of chess. And it was not only the technical calculations which led to that symbolic victory.

Deep Blue had been deliberately engineered not only to 'think' strategically but also to take on a human opponent.[50] As such, the machine sometimes paused, appearing hesitant and uncertain. This made Karparov complacent and over-confident in his own strategy. But it was the machine which eventually won, and it did so within the confines of unam-biguous objective rules.

Debating is subjective.

Miss Debater is perhaps even more capable of taking on the human condition than Deep Blue. She is able to turn a phrase, gently mock and respond to the vulnerabilities in her opponent's position. Not only that, she has the world of knowledge at her virtual fingertips. So the question is left hanging: did Harish win because he was 'right' or because he was able to deploy emotion understandably better to the human voting population? In these circumstances, were people ever going to pick machine over man? And just how important is that human connection when making a convinc-ing case for otherwise dry public policy?

Just like Amazon's recruitment technology, the Tec Tec suicide app, China's Social Credit and more, the issue here is human values, which are set to stay with us regardless of the power of the digital revolution.

IV

Imagine you were debating the future direction of the country and the issues in question surrounded the extent of public spending; just how many billions of dollars of tax payers' money would be required? If you were stating the case for, say, relative frugality, would it make a difference if your opponent had spent the morning relaxing in the late September sun?

It is perhaps a strange question. Your immediate answer is probably 'of course not'. But then you have been reading this book and by now appreciate just how much little things can have a lasting impact and that people's behaviour is not always rational. You will have considered that people voted for platforms proposed by Donald Trump and Jacob Rees-Mogg when objectively it was not in their interest. You will have observed that Harish Natarajan defeated Miss Debater by deploying emotion, not simply evidence. That is, the facts were not necessarily the most powerful part of Trump, Rees-Mogg or Natarajan's respective strategies. We also need to consider the nature of politics and politicians. It is rarely a 'truth-seeking' activity where actors attempt to dispassion-ately analyse given conditions to identify an or the answer. That is not the function of politicians or political parties. Connecting with others is what politicians do. Politics is a very human activity. And so is philosophy and values.

Let us leave the futuristic Miss Debater and go back to *the* broadcast event of 1960. Televisions were smaller back then. They were black and white of course. And the picture was less clear, more hazy. But that was no barrier to millions of Americans tuning in to watch the first Presidential debate between Democratic Senator John F. Kennedy and Republican Vice President Richard M. Nixon.

Moderated by the journalist Howard K. Smith, the can-didates were afforded an eight-minute opening and a three-minute closing statement. In between they would answer questions from correspondents and try to convince the unseen voters across the country to turn out for them. There was no studio audience, and so viewers and listeners at home had few cues as to what others thought as the debate unfolded.

One of the most frequently repeated observations about this debate is that television viewers largely believed that Kennedy had won, while those listening in on the radio thought that Nixon had taken it. Whether this is true is actually hard to determine to this day. But watching it back six decades later, it is difficult to dispute the view that Kennedy looked more confident, more relaxed, more assured in his communication. Nixon, by contrast, who had recently been in hospital, who had insisted on campaigning relentlessly right up until near his arrival in the studio, and who refused to wear make-up, looked a little gaunt, a little tired, a little uneasy, a little shifty even. It is also difficult not to consider that perhaps Nixon's content had just a bit more depth.

It is said that Kennedy's running mate, Lyndon Johnson, who had listened in on the radio, himself thought Nixon had edged it, while Henry Cabot Lodge, who shared the ticket with Nixon and who had watched on television, reportedly exclaimed that the 'son of a bitch just lost us the election'.[51]

This first ever presidential debate, and one which was broadcast to 65 million Americans, is seen as a real turning point in the campaign. The passage of time, of course, makes the detail of *what* was said much less relevant than in 1960, but even so, the real power of this great event was the theatre rather than the detail. As they vied for the highest office, it was the opportunity to watch these leaders up close to see how they conducted themselves, how they reacted to pressure and how they connected.

The idea that these small signals, assuredness and outward confidence, make a difference to how convincing a politician is when explaining a policy promise and therefore what we think about it, is extraordinary. There is research that shows better-looking candidates are more likely to win and experiments where audiences can predict the winner of an election by looking at photographs of the candidates.[52] We are superficial at times.

Rather than looking pale and fatigued with a five-o'clock shadow, as befell Nixon, the junior Senator for Massachusetts looked healthy and vibrant. This was not an accident and has something to do with how he prepared. Ted Sorensen, the future President's speechwriter, picks up the story: 'Kennedy — who had no debate coach and almost never rehearsed,' he remembered, 'arrived in Chicago the day before the debate and, after a long morning reviewing potential questions and issues in the sunlight on his hotel roof (his TV tan, contrary to reports, was not from campaigning in California), was sufficiently relaxed to nap.'[53] When he arrived at the studio, Kennedy was as calm as could be, and he not only felt it, his tanned complexion radiated that message to millions of voters.

And of course, there will be those who will say that Kennedy won on the strength of his argument, not the style of his hair, the cut of his suit or the tone of his tan. But then we have to remember that it was Kennedy himself who decided that the time invested sitting on the roof absorbing the sun's rays was worthwhile. And we need to remember that we are still talking about this after more than sixty years. The tan did not win the debate. That would be absurd. But as part of the package – content, delivery, style, emotion, that human connection – it gave John F. Kennedy an edge.

Perhaps, something like the experience of Deep Blue, politics is a game that must be played.

V

When we think about political debate in this way, the implications of a technology like Miss Debater opens up a world of possibilities. Just imagine a system that can debate any topic at all, one that is able to select and understand all evidence that supports the case and that which does not. One which also develops personality and a really meaningful and compelling narrative, one that responds forcefully to opponents. This technology could transform the way (democratic) public debate is experienced. If Kennedy had been debating a machine, would his tan have mattered so much?

And if that were the impact on more rational choice political discourse, would it have represented any kind of remedy to the populist and fantasy politics of more recent years? Would voters have trusted the factual rebuttals of Miss Debater when faced with a Donald Trump energising an angry base or Brexit campaigners giving an outlet to those left behind by globalisation? Trump managed to dismiss evidence contrary to his rhetoric with the simple but effective 'fake news'. During the UK EU referendum, Leave campaigner Michael Gove handled some inconvenient analysis about his position with the equally simple: 'I think we've all had enough of experts, haven't we?' Remember the reason why Joseph Franklin's suicide app fell short of its potential was the human processing of its analysis – users did not accept what it was telling them. It is all very well having powerful political technologies, but if the connection is human and evidence can just as easily be rejected in favour of something popular, there is a question about whether it can make any difference at all.

But wait a moment. It won't be one human candidate against a machine will it? Voters will not any time soon be electing a machine over a human. The power here is checking out the rhetoric and testing the substance. And that has implications for fantasy populists and the old rational choicers alike. And it starts with how we handle evidence.

VI

When British Prime Minister Tony Blair came to office in 1997, his government made a point of embracing what is known as evidence-based policy-making. It was stated firmly in his personal message at the top of New Labour's election manifesto: 'What counts is what works. The objectives are radical. The means will be modern. This is our contract with the people.'[54] And what this surely meant was that in 'doing what works', it is the ends that matter more than the means. That decisions should be driven by data rather than emotion. That the experts matter more than the ideologues. It allowed his government to accept globalisation and its implications with some alacrity, choosing to manage its challenges rather than join Corbyn and his cabal in a futile campaign against an unstoppable force. It was an approach that was intended to be inclusive in terms of bringing together those who just wanted things to 'get better' and deliberately differentiated Blair and his party from the dogma of the Thatcher years, which all so often had prioritised means over ends and divided people in her own party into two camps: the wets and the dries; those that were 'one of us' and those that were not.

Margaret Thatcher's long period in power was characterised by what political scientist Peter Hall described as 'Third Order Policy Change'.[55] That is, it represented a paradigm shift in the way we do things or think. Governments like this are naturally rare and have a lasting impact on successor administrations, regardless of party colour. Indeed, that third order can be thought of as causing a fundamental change in the way that the political establishment as a whole – government and opposition parties – perceive and think about politics. And that was only possible in the 1980s because the administration was so committed to its guiding philosophical ideals.

There is a story that during a policy meeting involving those who were not, as she would come to refer to them, 'one of us', Thatcher fished around in her handbag for a copy of Hayek's *Constitution of Liberty*, slamming it on the table with the exclamation 'This is what we believe!'.

New Labour was undoubtedly a product of Thatcherism and the third order policy change her government represented. Nevertheless, Blair's social democratic politics were much less ideologically rooted, and he teased the prospect of policy change informed by intelligence, experimentation, data, analysis and reason: in other words, evidence.

It all sounds rather antiseptic. But was it?

VII

A conclusion could be drawn from this that professional politics and rational choice has supressed ideology, and that is the singular issue for tackling the major challenges facing our societies. But it is not as simple as that. It is not that ideology is supressed, but rather it is either reframed in scientific language or we are in the Fukuyama territory of end of history or Sorman's 'economics doesn't lie'. But there is another risk in a retreat to philosophical beliefs and it is this: tribal ideology alone presents the danger of delivering fantasy politics.

Back in 2010, the 10th edition World Social Forum returned to Porto Alegre, the largest city in the southern region of Brazil. Porto Alegre had hosted the first three editions of this annual meeting which brought together civil society and non-governmental social organisations as well as campaigners and solidarity movements from across the world. WSF delegates are interested in alternative globalisation, campaigning for social justice and coordinating strategies. The WSF, then, represents something of a counterpoint to the World Economic Forum, which meets in Davos, Switzerland, each year at around the same time, and which might be regarded as a capitalist organisation in which leading economic policymakers, investors, business and thinkers shape the global agenda. It has been at the forefront of understanding globalisation and indeed the Fourth Industrial Revolution.

In 2010, The WEF's ambition was to conceive of sustainable recovery. The world had been convalescing from the greatest challenge to globalisation and the longest and deepest economic contraction since the Great Depression itself. The Financial Crisis had shaken global economies just as it had squashed Fukuyama's vision. There were tensions on open display between policymakers who had authorised huge interventions into the global economy and investment bankers who had been blamed for the crash. French President Nicolas Sarkozy set the tone arguing that 'we can only save capitalism by rebuilding it, by restoring its moral dimension'.[56] Nevertheless, it was clear to all that the show must be kept on the road.

Over on the other side of the planet in the warmer climes of Porto Alegre, it might have actually been a different world. Indeed, WSF ambitions in contrast to their WEF rivals were 'challenges and proposals for another possible world'. Delegates at the WSF were less interested in better behaviour to make capitalism work than replacing capitalism altogether. And it is instructive to see the themes which had gathered into ideological creed over the Forum's first decade and been compounded in the years since.

The WSF adopted a Charter after its first meeting which envisaged interconnected civil society groups creating alternatives to neoliberal capitalism and believing in 'human rights, the practices of real democracy, participatory democracy, peaceful relations, in equality and solidarity, among people, ethnicities, genders and peoples.'[57] Set in opposition to globalisation, there is some consensus across WSF delegates about rejecting liberal, free markets; about limiting capital; about protest; about environmentalism.

And yet there is no real reason the likes of environmentalism should be the ideological preserve of the left or freedom owned by the right. Being free should reasonably be a priority for the left of democratic politics. That idea of stewardship should be equally at home with conservatives. Alas, that is not how ideological positions have hardened. What we see is also how political positions have emerged really since the 1960s. The left's commitment to environmentalism reflects its ideological aversion to corporations and economic freedom and the political right seeing that environmentalism as a challenge to its own economic interests. Freedom is seen by the right as an economic freedom for the wealthy while the left is suspicious of allowing individuals to make decisions that are not good for them or the collective. But it does not always stand to reason.

It is what the US economist Jonathan Haidt is talking about when he says that morality 'binds and it binds'. As he puts it, 'Follow the sacredness, and around it is a ring of motivated ignorance. If you know what a group calls sacred you will know where they're blind to reason'.[58]

VIII

Imagine a world in which a machine, like Miss Debater, could assess all available evidence, cut through the lies and spin, rise above the sacredness that binds and form a dispassionate policy on the issues of the day. Imagine this combined with virtual twins, digital cities economies and the virtual planet DestinE, which can model and interact with the physical world, give feedback and offer certainty. It might be a bit like a shiny robotic Tony Blair from 1997 (before he waged war in Iraq on the flimsiest of evidence at all) wrapped up with statisticians, policy wonks and filing cabinets full of policy in action reports from every corner of the world. Imagine how transformative that might be. How it might make redundant much of the frustrating pantomime of democratic politics.

But that would also be a mistake. To pretend there is no ideology is to reduce issues to a question of correct and incorrect and management efficiency. It suggests that evidence does not require the lens of political values through which to interpret and to prioritise. There is rarely any such thing as a single evidence-based truth. And both rational choice politics and disruptive fantasy populism have been denying that truth for years. Furthermore, recall how even knowing that Trump was misrepresenting reality did not change how supporters felt about him; there is plenty of evidence from psychological studies that facts alone are insufficient to motivate a political decision.[59] Some of this was on display in Porto Alegre just as it can be found in Davos.

The problem too with ends justifying means is that if the evidence leads to an unpopular policy, politicians avoid taking the decisions. That is what happens in rational choice politics with difficult, long-term issues. Take an issue like climate change. The (environmental) evidence would overwhelmingly point to policies which reduce incentives to emit carbon, but that means curbing citizen's ability to take cheap flights or drive their cars through higher taxes or regulations. When energy prices rise, as they did so dramatically in 2022, there is an inevitable call for policies to be ditched. And the benefits of the measures? They would likely accrue to a future generation – not the one paying the bill today. It is easy to see how that is a hard political sell. And it is even harder when opposition politicians can promise to tackle the problem in some illusory way that will not cost the voter (regardless of whether it can be delivered).

The technology has the potential to transform the way we understand and deliver politics. But only if we combine the evidence with our values and are willing to be challenged on what we believe we hold sacred.

IX

Mohamed Bouazizi was a twenty-six-year-old Tunisian street vendor who, on 17th December 2010, set himself alight in public, protesting at the harassment he had received at the hands of officials. His extreme act was to be the catalyst for a movement which quickly spread through the country.

Unemployment, food poverty and freedom were at the heart of the demonstrations which clashed violently with authorities. The 'Jasmine Revolution' as it became known, defied government crackdown and eventually overwhelmed police and military. Just a month later, on 14th January 2011, President Zine al-Abidine Ben Ali, who had held power since 1987, fled Tunisia.

The first of the pro-democracy uprisings happened in Tunisia and then Egypt, when protestors toppled longstanding autocratic regimes. This led to a wave of similar protests in North Africa and the Middle East. The Arab Spring saw tensions erupt about economic conditions, human rights, freedom, dignity. Bahrain then Libya then Syria and Yemen, Sudan, Morocco, Oman, Iraq, Algeria and other countries in the Arab world saw uprising, violence and varying degrees of political and social change. As Klaus Schwab put it, introducing that World Economic Forum in 2011, 'Global youth is rising to the challenge of leadership in the twenty-first century using new technologies and ways of communicating. New actors are emerging and influencing global events in unanticipated ways.'[60]

What was extraordinary about the period was something about the cause and effect. Just how did the bold defiance of a young man who sold vegetables from a barrow in the anonymous town of Sidi Bouzid lead to the incredible mass revolts which were to reverberate right across the region?

There was one very powerful feature of the Arab Spring that quickly became apparent. The relatively new technology of social media not only facilitated communication between activists, across protests and to the outside world, helped organise activities, mobilise people, shape debate and disseminate ideas, but it also seemed to be something of a match with the demands for freedom at the heart of the uprisings. This new technology, some believed, empowered citizens in a truly democratic way which authorities, so used to control, were unable to quell.

Social media had this potential power to open up information, resist control and disseminate ideas freely. It has the potential to offer some sort of mass direct democracy. And for democracy to flourish there needs to be openness so that power can be challenged, voices heard and new ideas emerge. But these were the days of optimism for not only liberal democracy but also the possibilities of technology in breaking down barriers to communication and debate as we head towards the inflection point of digital transformation.

In the decade to follow, we became more cynical about the power of social media which poses threats to as well as opportunities for democracy. It must be recognised that the business models of the tech giants which own global social media are not driven by these ideals of providing open direct democracy but rather to harvest huge data and provide content which engages attention. This goes some way to explain why the 2021 US Congress hearing – which called on testimonies from Mark Zuckerberg, Jack Dorsey and Sundar Pichai – chose the title: 'Disinformation Nation: Social Media's Role In Promoting Extremism And Misinformation'. And it was Ilinois Congressman Robin Kelly who made the case that 'To build that engagement, social media platforms amplify content that gets attention. That can be cat videos or vacation pictures, but too often it means content that's incendiary, contains conspiracy theories or violence. Algorithms on the platforms can actively funnel users from the mainstream to the fringe, subjecting users to more extreme content, all to maintain user engagement.'[61]

The result of that is the risk that social media narrows down debate rather than opening it up. It risks blinding and binding. Cass Sunstein, for instance, has warned of the perils of polarisation that results from like-minded people who not only congregate virtually but also reinforce their pre-held views through the consumption of the same information. Those algorithms, far from opening up information, actually filter and isolate in an 'information cocoon'.[62] And while there is some promising evidence of cross-ideological dissemination,[63] social media today too often serves as an echo chamber feeding the appetite for fantasy politics, rather than as a tool for empowering citizens to make informed decisions.

Combine that observation of how political social technology has tended to work with revelations from human psychology and a greater sense of the challenge emerges. Frimer, Skitka and Motyl wanted to discover just how ideologically open-minded people were, so they conducted an experiment. Suppose you were given a choice: on a given political issue, you could read an opinion you agreed with or one you did not. The issue might be same-sex marriage, elections, drugs, guns, abortion. If you read the opinion you agreed with, you could win $7. But, to make it more interesting if you picked the opposing viewpoint, you could win $10. What would you do?

You might not be surprised to learn that almost two thirds of participants decided they would sacrifice the potential extra cash rather than be exposed to a viewpoint they disagreed with. This was true of liberals and conservatives alike.

And the aversion was not because participants felt they were already knowledgeable on the subject. Often they were not. The explanation, these researchers concluded, was that people want to avoid cognitive dissonance, that is the anguish or frustration that comes from holding conflicting views.[64] We do not want our views or prejudices challenged, we become quickly tribal and we dislike hearing different viewpoints. Moreover, mediums such as Twitter magnify the illusory truth effect as simple statements are communicated quickly and often without the space for nuance. It limits 'earnest discourse'.[65]

While technology does indeed offer increasing opportunities for public engagement in issues and policy, it is important not to overplay its impact so far. In his book *Beyond Slacktivism*, the academic James Dennis even argues that easy use of online political acts, such as signing an e-petition, not only have limited impact but might even be dangerous, since these acts lead to a sort of fulfilment which disincentivises people from actually getting involved in politics or genuine activism.[66]

The possibilities of the technology as it continues to develop into augmented reality and the metaverse are profound and are set to change the way we access information, news, and communicate; how we work and interact; what we think is important and determining public priorities.

The technology alone, whether in public or private hands, will never be a sufficient precondition to strengthen politics or prepare our democracies for the Fourth Industrial Revolution. Just as in the workplace, the true power of the technology can only be harnessed in collaboration with humanity. And once again it is a lesson that the revolution must be reclaimed or else it will engulf political discourse.

XI

For Larry Fink, the billionaire boss of investment firm Blackrock, the Russian invasion of Ukraine in 2022 not only represented an 'humanitarian tragedy' but actually 'put an end to the globalization we have experienced over the last three decades'. In a forthright letter to shareholders, he explained how the sanctions imposed by the West demonstrated the strength of capital markets, but that the easy global expansion of trade, markets and growth which we have come to accept, was now being challenged. 'Russia's aggression in Ukraine and its subsequent decoupling from the global economy is going to prompt companies and governments worldwide to re-evaluate their dependencies and re-analyze their manufacturing and assembly footprints – something that Covid had already spurred many to start doing.'[67]

It is a stark warning that underlines the political challenge of this inflection point. Globalisation has led to economic prosperity but has also created large numbers of dissatisfied, 'left behind' citizens. That in turn has fed a fantasy politics and populist movements across the industrialised world. The real wider political consequence of the Ukraine war could well represent something of a blunt weight that drags on the momentum of globalisation; a force that was already reaching its sell by date as the digital revolution approaches.

But the Russian invasion also revealed something else about European politics in particular, and it is back to that important point about values. Brexit, like so much fantasy politics, was about division. It was about dividing society, exploiting discontent and espousing certainty of the future with easy popular messages – one of which was the idea of some sort of buccaneering global Britain. Russian aggression blew a Brexit-shaped hole in Boris Johnson and Jacob Rees-Mogg's rhetoric. The crisis divided the world into the democracies who were viscerally opposed to Russia and those (mainly) totalitarian regimes who were unconcerned. In a world that was becoming even more complex, there was clarity over national values and interests. Ukraine demonstrated not only how essential it is that powers like Britain work together with their European partners, but more fundamentally that these are partners because of shared fundamental values, not just geography. That populist idea of Britain forging new deals with anyone and everyone around the world regardless of values was always a fantasy and the geopolitics of the war in Ukraine underlined that. It solidified the resolve of the EU (and NATO) on defence, energy, trade and, yes, values. Moreover, it put them in stark opposition to the non-free world headed by Russia and China. Remember, these are also the rival ideologies that will clash and shape the Fourth Industrial Revolution.

The political environment then is nuanced, complex and cannot be addressed with easy slogans and binary statements. If globalisation is eventually making way for a new force, the digital revolution promising to reshape our world, and geopolitical threats meaning a re-evaluation of dependencies, domestic politics must be reclaimed and not simply by the usual elites who have only been able to pursue aggregate policymaking.

XII

Eupen-Malmedy is a small region in Eastern Belgium. It is predominantly German-speaking and is commonly known as Ostbelgien. It is a picturesque area and many people visit to walk and to cycle. Its history is long – it was once part of the Prussian Empire and is today one of the three federal communities and four language areas of Belgium. That is plenty of divisions for such a small country.

Like other industrialised economies, Belgium enjoyed the fruits of the golden age of capitalism and what some people called an 'economic miracle' in the aftermath of the Second World War. Back then, demand for the outputs from its industry soared. But this small economy has felt deindustri-alisation too and while it benefitted from being the centre of EU power, the unskilled continued to have felt left behind. And all that means there are issues which day-to-day politics struggles to properly address.

In 2019 the Ostbelgien Parliament legislated to create its very own Citizens' Council to work alongside the existing elected chamber. Twenty-four citizens, selected through a lottery, rotate every eighteen months and deliberate over issues and problems, making recommendations to the Parliament – which in turn is obligated by law to debate – at least twice – and reply.

Think of the possibilities of this. It means that citizens not only have a strengthened voice in politics, but they also take on more responsibility for the decisions made. They are neither voting themselves rich nor enjoying the deceptive comfort of popular fantasy politics.

What is happening in Ostbelgien is limited for sure: it is non-binding, it does not really set its own agenda, it is far from ideal. But it is a start. And it is an example of 'delibera-tive democracy'.

Back in 2003, Chambers wrote a brilliant review of delib-erative democracy, showing that it comprises broad debates and different definitions.[68] The early work was as theoretical as it was idealised,[69] but still it focussed on a hope that political outcomes could 'secure broader support, respond more effectively to reflectively held interests of participants, and generally prove more rational'.[70] That is surely what the proponents of the Citizens' Council in Ostbelgien wanted.

And that practicality is what interests more recent thinkers who write about systemic and empirical assessments of deliberative democracy.[71] They are interested in setting those ideas in the real world and in real experiences and systems – experiences like those in Eupen-Malmedy. This all advocates moving away from rational choice or indeed fantasy populism to a situation where 'citizens make political choices freely, following extensive debate and discussion regarding the implications and consequences of those choices, both for themselves as individuals and for the society as a whole'.[72] Note that well – citizens take responsibility for political and economic choices. So, we have what Mansbridge calls 'citizen-centred' democracy and what Stevenson and Dryzek describe as 'authentic, inclusive and consequential'.[73]

XIII

Porto Alegre is a very different sort of place to Eupen-Malmedy. This Brazilian city, which hosted so many World Social Forum meetings, is the cosmopolitan capital of the country's Rio Grande do Sol state. It is home to some one and a half million people and, located at the juncture of five rivers, is a significant industrial centre and port. But Porto Alegre is known for something else and that is that it was the first city to implement what is known as 'Participatory Budgeting', its practices dating back to 1989.

Participatory Budgeting is a process where ordinary citizens are asked to allocate public spending and to determine public priorities. Public Administration Professor, Celina Souza, has documented the experiences in Porto Alegre, and the insights are truly enlightening.[74]

Brazil is a country which has not only experienced re-democratisation but also suffers from 'deep-rooted' economic, social and political disparities. This makes the problems of aggregated policymaking all the more stark, and Brazil has tried to 'reinvent government' and decentralise power. In Porto Alegre, there are plenary assemblies which meet in various parts of the city and whose job it is to help write the budget. Officials present information but citizens determine the priorities. Once this is done members are elected by each district to form the city-wide municipal council. The decisions are not always the raciest, with preferences in Porto Alegre tending to direct resources to street paving, sewers and housing. The broad thematic areas are transport and traffic, education, leisure and culture, health and social welfare, economic development and taxation, city organisation and urban development. So it can be seen that citizens' responsibilities are wide and potentially strategic. Causation or correlation, make of it what you will, but citizens of this city have notably high political awareness and perhaps crucially communal trust.

Participatory Budgeting can be seen to have been a long-term success story in Porto Alegre. It suffers from the 'implementation problem', of course, where it is often difficult to wrest decision making out of the hands of the most powerful and to engage those from lower socio-economic groups. Nonetheless, the process is deliberative and allows citizens to truly influence the political decisions taken in their own local economies. They have to make difficult decisions, they need to apply their values, they need to collaborate and cooperate and they need to share responsibility.

XIV

You will see where this argument is going. It is not enough to recognise the folly of aggregate policy making and strive for inclusive growth any more than it is enough to experiment with letting citizens make some decisions. Just think about the power of the pooled virtues of inclusive growth and deliberative democracy as we approach Industry 4.0. There is so much crossover in the ideas, and yet they have remained in their theoretical and practical boxes. Combining them offers a way to draw citizens into the decision-making process and transform economic opportunities. The transformative potential of this human, political approach would not simply magnify the possibilities of emerging political technologies, but would actually give those technologies a political purpose which serves citizens' needs. This is the sort of collaboration that could prepare societies, economies and communities for the disruption of the Fourth Industrial Revolution.

Of course, this all threatens the existing order – the elites who continue to dominate – and it threatens it in a more permanent way than populism. Indeed, it threatens rational choice and populism alike. And that is a good thing. It is not class war, soundbites, empty promises or fantasy that will address the great issues in the Fourth Industrial Revolution. It will be highly human skilled open-minded communities, freedom, genuine inclusive growth and deliberative politics.

And that leaves a big question. When we look over the

precipice into Industry 4.0, will the technological revolution strengthen our democracy or weaken it? Will our politics be strong enough to shape the Revolution or simply react to it in the predictable way it has reacted to change for decades?

One very effective way that politicians traditionally tackle the trickiest and most wicked of problems is to set up some sort of commission or inquiry that allows 'experts', supposedly unincumbered by politicians, to investigate. Imagine we were able to do this with technology and rather than experts and pseudo-science, it was genuinely deliberative and inclusive. That is there is no pretence that an issue is purely logical, factual, non-political. It means that we all take responsibility for ambiguous and difficult policy.

The likes of Miss Debater could become a powerful tool in navigating communities through this inflection point and the most difficult of decisions. Politics would become no easier, but it would be a step away from rational choice and fantasy populism that pretends issues are simpler than they really are and that unpopular choices can be avoided. Remember Miss Debater was allocated a side of the debate when she challenged Harish Natarajan. The possibilities here are that the technology can help us to understand the perspectives of others – collaborate and to make better decisions. But those decisions need a real sense of shared values. After all Social Credit tackled an identified problem and might even be said to be effective. But it was far from being apolitical and reflects very distinct authoritarian values.

Technology can help us to make better decisions and can strengthen democratic politics, but only if we drop the fiction that there can be pure scientific evidence. Like with the earlier debates, the technology needs our humanity and a strong sense of values. The revolution before us is digital, but it will free up human capacity. Politics needs to decide what we do with that capacity.

Politics as usual has failed to resolve so many of the big questions which have been with us since industrialisation. Politics as usual shows little ability to get to grips with the big challenges of our time. It remains driven by aggregated measures not disaggregated experiences. The problem is at least partly systemic, as democratic practices not only punish politicians who make the toughest of decisions but also disincentivise inclusivity. Too many have been left behind by globalisation, taken for granted by established parties and entrenched positions. The rebellion has been to propel fantasy politics to the fore, something so counterproductive when the Fourth Industrial Revolution is on the horizon. The prize is there for progressive politicians to grasp the challenge of this inflection point, to acknowledge that so many communities are unprepared for what comes next. To make sense and to combine inclusive economics with deliberative democracy. It is to embrace an openness to ideas and approaches that is the political equivalent to Building 20, or lots of them.

To prosper in tomorrow's economy, the skills that are desperately needed are creativity, adaptability and leadership. We need to be able to harness the human resources available to make a success of the Fourth Industrial Revolution and make growth inclusive. Politics needs to be reclaimed, the political technology needs to work for humanity, supporting deliberative government. It will only happen with great leadership, and our experience of that over many decades is questionable.

PART 3:

EXTRAORDINARY ADVENTURES IN LEADERSHIP

How the Tragedy of Grenfell Shows that Great Leadership is Rare... But Can Be Found Everywhere

'We've done more in three hours than the government in three days'

I

'We were making as much noise as we could outside to wake people up, but we woke them up to die. I wonder now if they might have been better left in their beds.'[1]

These chilling words were spoken by fifty-two-year-old Alan Kempthorne, a member of the public who had spent the early hours of 14th June 2017 trying to warn residents in the Grenfell Tower block in London's Kensington that it was on fire. That fire would rage through the twenty-four stories, killing occupants of twenty-three of the 129 apartments, providing a desperate imagery that would shock a nation and raise questions about societal priorities. It also represented a dramatic study in leadership from those at the very top who had sought power to ordinary citizens unexpectedly confronted with a crisis.

Grenfell Tower was completed in 1974 and designed in the brutalist style common to the architecture of those decades after the Second World War, as towns and cities were rebuilt, the state expanded and the demand for housing met. Unveiled as modern and serious, often such constructions are today associated with urban decay and depravation. And by way of softening their raw, cold, uncomfortable appearance, many have undergone cosmetic retouching such as pebble dashing and stucco.

Grenfell was such a building. In the Borough Council's own documents it is described as 'designed as a large rectilinear mass lifted high off the ground on stilt-like columns and nestled in an urban garden'.[2] The concrete structure stretched almost seventy metres into the sky in uniform rows of mill-finished aluminium windows interspersed with monotonous dark materials encased in an unrelenting cuboid form. It was renovated after 2012 to include new windows and zinc composite cladding intended to offer 'a clean appearance, crisp detailing at joints and an attractive dull lustre'.[3]

From the beginning, Grenfell was home to some of the poorest in the area – a wider community of Kensington which in 2017 it should be noted was Britain's wealthiest and which contained some of the richest people in the world. Once known as 'Moroccan Tower', Grenfell was still home to scores of families with ties to Morocco at the time of the fire. Like other economic migrants, many Moroccans were attracted to London in the 1960s and since in search of work in catering and the hotel business.[4] When the victims of the fire were named it was clear how multicultural Grenfell was with immigrant communities mixed with those from British heritage families; all ages represented. And while a few of the residents had become home owners, what also became clear was how forgotten these poor people had become.

Those who survived that terrible night escaped with their lives but with very little else. The fire engulfed 151 homes and everything in them. People were instantly homeless, suicide attempts in the area rose, and the community was in shock, with anger spilling out in front of the many television news cameras to cover the story. They wanted to know why this had been allowed to happen. They wanted to know why they had been overlooked. They wanted to know what actions were going to be taken. It was a situation that called not just for competent management of emergency services, health and social responses. It was a situation that called for leadership.

II

Leadership is ethereal. It is so often positional, associated with a post or role, but has more to do with behaviour, foresight and presence. Human society yearns for leadership and leaders influence so that the group can achieve a common goal. This inflection point is no different.

Leadership has become a big subject in academia, professional life, business, politics, science and sports. Warren Bennis for instance has explored the qualities that make for a great leader,[5] while Maxwell has identified the need for leaders to 'shift' to adapt rapidly.[6] But perhaps in simple terms, whatever the categorisation or changing definition, we know leadership when we see it. It is constructed from society, inherently contextual and the product of the observer. And leadership has always been necessary for human survival. It is in us as human societies.

Leadership is when collective action is needed and societal norms of cooperation are insufficient to solve the problem. But our leadership problem is that all too often the leadership we experience is simply not good enough. It is too incompetent, too dictatorial, too self-serving – as the previous section illustrated. We get popular fantasists when we need insight and deliberation. All the unsolved problems discussed in this book, all the people forgotten or left behind, all the debates unresolved is the consequence of this. Leadership is not good enough today and it is not good enough for the Fourth Industrial Revolution.

Like the economic and political debates, leadership has both evolved and simultaneously finds itself stuck in poor and counterproductive practices. Each previous industrial revolution has been characterised to some extent by a style of leadership. You can imagine what they are. At the beginning the hero leader charismatically inspiring followers to see the possibilities in new inventions and processes. By the second industrial revolution and the introduction of electricity into the established factories and firms, 'scientific' management, command control approaches maximised the productivity of the human capital. Then in the twentieth century, as computer technology developed, there was the idea of moving beyond these transactional relationships with leaders inspiring their followers to 'transform'. And with all of this came the ambition to performance-manage.

Anyone looking round leadership today – at politicians, CEOs, managers, public figures – will not fail to notice that there are still charismatic leaders, dictatorial leaders, good leaders, poor leaders, indifferent leaders, misguided leaders. There are good leaders who exercise poor judgement and bad leaders who somehow make the right decisions. There are formal and informal leaders, there are the powerful and the weak. There are plenty of leaders who fail to lead, plenty of leaders who demand compliance when autonomy would produce better outcomes, plenty of leaders whose actions and rhetoric are inconsistent.

Leadership has not been some kind of linear progression through industrialisation that now means a new regeneration for the Fourth Industrial Revolution. But the inflection point does represent real challenges for leaders and demands particular skills, traits and ways of approaching the world. This chapter unravels the idea of leadership and what it means today. It shows behaviours that still stand for leadership despite being counterproductive, regressive or even corrosive. Given what we know about this inflection point and the challenges to us all, we need to be able to make sense, to collaborate and to harness the potential of our societies. Leadership, after all, is not necessarily about position in the hierarchy but is more about influence, vision and empowerment.

If nothing else, the chapter makes one enduring point: leadership is something that must be reclaimed. And that is a tall order!

III

As dawn broke over Kensington that June morning and thick smoke hung in the air, television news crews swarmed over the devastation. People were already asking 'why?'. And what unfolded at Grenfell was a failure in leadership from those in charge. The local authority, Kensington and Chelsea Council, was incapable of handling the crisis and appeared indifferent to the plight of those affected. Meanwhile, the aloof image of the Prime Minister, Theresa May, and the embattled government she led was only underlined by the mistaken decision to refuse to meet victims. And what of the man of the people, Jacob Rees-Mogg? Well, he took to the airwaves the following Monday telling LBC Radio: 'The more one's read over the weekend about the report and about the chances of people surviving, if you just ignore what you're told and leave you are so much safer. And I think if either of us were in a fire, whatever the fire brigade said, we would leave the burning building. It just seems the common sense thing to do.'[7] It was an ill-advised comment which commanded a profound apology the next day.

The failure of leadership was historic, in that the problem was not addressed before catastrophe struck, as it should have been. It was a failure to prioritise those that were not powerful and whose voices were not heard. But it was patently a failure of leadership in response.

IV

How leaders respond to or in crisis is as instructive as it is important. A crisis is when we need leaders to make sense of what is happening, to reassure and to galvanise. After all, words are easy when times are easy. Being inspirational is nowhere near as difficult on the up as it is to use failure as a springboard into success. Leaders need to be judged by their actions. And in a way that is part of the problem. There is a myriad of reasons why individuals are able to assume leadership positions. That they do and say the 'right' things is one of them (or in Theresa May's case she found herself as the last remaining 'grown up' in the room).[8] The promise of progress and good times, accompanied by the overused slogans of being 'on your side' or that 'we are all in this together' count among the most compelling pitches for position even if they are not always a reliable indication of performance. You will have heard them from politicians and bosses alike. No doubt most leaders are not lying when they talk about their values and they articulate the sort of principles to which followers can relate. But how many really can live up to those values when the chips are down? How many are being disingenuous (even to themselves)? How many are deluded about their impact? So many leaders disappoint us, not simply because of their incompetence but because what they do is at odds with what they have said. It is possible to respect someone with whom we disagree and even to remain a (loose) follower where they are principled and consistent. But the most powerful leadership-follower relationship is when there is a shared vision, a shared philosophy, a shared understanding of what the leader wants to achieve with their followers and, to some extent, how they will all do it. It is in difficult times that real leaders rise to the challenge and support those for whom they are responsible. But that mettle is not always tested until the difficult time is upon us and sometimes that is too late.

There is a darker explanation too. There is an extended literature on the corporate psychopath who is over-represented in boardrooms. These are people with sociopathic traits, a lack of empathy and remorse. But it turns out that their ascent is also faster. They get promoted quicker than the rest of us. That is, the traits - the ruthlessness, the selfishness, the charm – are still valued in organisational life. And they are too frequently valued over thoughtfulness, deliberation and openness. It goes some way to explain how populist figures unsuited for leadership, manage to rise.[9] It is but one explanation as to why we have so many unsuitable leaders.

So often, where leaders lose the confidence of their followers is when they break the bond through their actions which contradict followers' understanding of what they are (or were) about. In essence, they force followers to disengage their allegiances. Think of all the organisations with their carefully crafted mission statements and values, universally positive and so often employing catchy acronyms. These values have to apply to bosses as well as employees. It will come as little surprise that followers notice when their leaders do not live up to the values they have printed in glossy brochures or plastered over their intranets which they expect all personnel to embrace. And what of those values? Eric Anderson and Brad Jamison studied the values statements of the biggest US companies. It is astonishing just how similar were the choice of words selected. The top ten were Integrity, Respect, Teamwork, Innovation, Quality, Performance, Excellence, Trust, Diversity, Leadership.[10] Sound familiar? They are almost so universally agreed as to be uncontroversial and near meaningless. They give little sense of what the organisation is about and how its community works together. There is no story that is any different from every other company. And as is evident from the adventures in this book, the digital revolution before us is a challenge for human values. It is a challenge to understand and to apply them.

That storytelling, the sense-making, is an important skill for any leader. In fact, we rely on our leaders to help us process change. We want them to offer some certainty in an uncertain world. We rely on them to articulate purpose. But it is only by trusting our leaders that we can handle the lack of certainty. And that requires leaders to trust those they lead. By goodness that is a crucial point to reflect on at this inflection point.

Trust is not something that comes naturally to those leaders who embrace the fantasy politics of populism. For them it is not the creative possibilities of collective psychological safety that drives their agendas but rather the stoking of conflict and division. That is not to say it is ineffective. Populist rhetoric can successfully feed the leader-follower dynamic, pitting supporters against perceived 'enemies'. While it might be effective in sustaining a populist leader, however, it is usually counterproductive and when one reflects on the needs of the Fourth Industrial Revolution, it is difficult to imagine populism as doing anything but cause damage and chaos. This digital revolution is complex, nuanced and uncertain. Politics as usual, leadership as usual, prefers simple solutions which are naturally more attractive. Psychologically we have a tendency towards what Taleb calls the 'narrative fallacy'.[11] That is, humans are attracted to simple chronicles manufactured to connect events logically and explain, even if they are not accurate. It is similar to the illusory truth effect of populism. Real leadership needs to embrace the complications.

Here is an interesting contrast in leadership. During the summer of 2021, populist British Prime Minister Boris Johnson decided to throw some red meat to his supporters with a rhetorical assault on criminals. No one likes crime, and tackling law-breaking is always to be found towards the top of voter concerns. But this did not represent anything like a policy agenda to reduce criminality or to make the streets safer. Appearing alongside his hard-line Home Secretary, Priti Patel, at Surrey Police headquarters, Johnson talked tough enough to create some headlines: 'If you are guilty of anti-social behaviour and you are sentenced to unpaid work, as many people are,' he pronounced, 'I don't see any reason why you shouldn't be out there in one of those fluorescent-jacketed chain gangs, visibly paying your debt to society.'[12]

Of course there was no real or substantive policy (even to roll out 'chain gangs') in the styled 'Beating Crime Plan'. And there was no attempt to understand the causes or impact of crime. The tough talk was enough though to create division, to mark out 'enemies' and to enrage the sort of 'chattering classes' Johnson had so successfully positioned himself politically to annoy. But it was a simple tweet from the chief executive of a familiar family firm which highlighted the inadequacies of this type of leadership.

'Instead of making offenders wear high-viz jackets in chain gangs, how about helping them get a real job instead?' asked James Timpson. 'In my shops we employ lots of ex-offenders and they wear a shirt and tie. Same people, different approach, a much better outcome.'

Timpson is a presence on the high street up and down Britain. It is where shoes get repaired, keys cut, clothes dry-cleaned and even broken mobile phones fixed. It can trace its history back to the nineteenth century and, with 2,000-plus outlets, boasts an annual revenue of more than £300 million. So it is a successful company, doing traditional work in a digital age.

James's father, John Timpson, had been CEO before him and pioneered what he calls 'Upside Down Management'. He explained this philosophy in an interview to *Management Today*:

'It's the way we run our business, started because the only way to give great service is to trust colleagues to look after each customer in the way they feel is best. So we only have two rules: 1. Look the part. 2. Put the money in the till. We got our first bonus from Upside Down Management in 1996, when I visited our Timpson shop in West Bromwich and discovered Glenn the manager was taking in a few watch repairs. He apologised, but I certainly didn't mind when he told me his watch repair sales were over £100 a week. It triggered our watch repair service, which Glenn has played a key role in developing into a £30m a year business.'[13]

So, upside down management is simply about trusting the people who work in the organisation to do their job in the way they think is best; handing over control. It is about the leadership supporting staff, not telling them what to do – that means a business without long lists of rules. And the hands-off style means creativity can emerge from anywhere as that first example of when Glenn started repairing watches. Not only that, but there is also a kindness to the leadership which demonstrates that those who work at Timpson are valued. Staff are paid bonuses, given their birthday off and can use the company's holiday homes. And if they get married, expect a week off work and free use of the company limousine. This might remind you of how the Richman brothers treated their employees back in the 1930s. It is little wonder that Timpson is an omnipresence in the *Sunday Times 100 Best Companies to Work,* featuring in the top ten every time it has entered.

James Timpson wholeheartedly embraced the leadership approach his father championed and under his stewardship the company has continued to grow. But James pioneered an approach of his own which took the idea of trust to a different level, and it is this that he was highlighting in that powerful tweet. Led by James, Timpson has become one of the UK's biggest employees of ex-offenders. More than 600 staff, around 10%, were once behind bars. Timpson's charitable foundation runs training academies in seven prisons, and its managers visit Category C and D prisons (which include those convicted of violent crime and theft but are either open allowing for day release or have a training and resettlement provision), to recruit new employees. 'It's a great way of finding amazing people' is James's explanation. 'Most companies employ ex-offenders but they just don't know it, as they had to lie on their application form to get a job.'[14] What a way to ensure potential is realised.

And this 'trust' model which contrasts so forcefully with the populism of Boris Johnson cannot be dismissed as some sort of failed left-wing communitarianism. John and James Timpson are capitalists. Their family business they own makes £20 million profit each year. Moreover, while the emphasis here is on the leaders trusting their followers, you can bet the staff at Timpson trust their leaders. And that could be becoming unusual.

VI

How much do you trust your leaders? Take your President or Prime Minister, elected representative, your manager, your boss – do you trust them (and do they trust you)?

Edelman has been running its Trust Barometer since the turn of the millennium. It shows the decline in trust we have for leaders in public life, business, journalism and, yes, experts. Inequality is seen as a driver of mistrust and lack of trust is more pronounced with those who are struggling to make ends meet. That is a problem for progress and a cheap opportunity for populists. One only needs to consider the vulnerable communities hit by crime that Johnson sought to enrage with his call to bring back 'chain gangs'. His intention was not to reassure but to divide; it was to exploit mistrust. In a conclusion familiar to the analysis of this book, Edelman's results show 'a world of two different trust realities. The informed public –wealthier, more educated, and frequent consumers of news – remain far more trusting of every institution than the mass population. In a majority of markets, less than half of the mass population trust their institutions to do what is right. There are now a record eight markets showing all-time-high gaps between the two audiences – an alarming trust inequality.'[15] So, we do not trust our leaders for two reasons that the Grenfell tragedy underlines: because too often leaders are not good enough and because of the divisions in our social fabric.

Citizens left behind by globalisation, those in Cleveland who might have once worked in factories like Richman Brothers or Nuneaton with its now demolished Courtauld's factory, understandably take a rather sceptical view of authority. And that permeates down. Meanwhile Gallup's State of the American Workforce survey confirms that workers feel underappreciated and most report that their leaders manage their performance in a way that does not motivate. That is perhaps because leaders are too often concerned more about (measurable) performance than what their teams are capable of doing. More of that in a moment.

And it is not simply about how happy a workforce is or retaining staff who might otherwise leave. They are more productive. In fact, according to fascinating research published in the *Harvard Business Review*, the neuroeconomist Paul Zak showed that 'high-trust' companies are 50% more productive than 'low-trust' companies. Employees have 13% fewer sick days and staff are 76% more engaged. They report 106% more energy (and 74% less stress).[16] Companies like PriceWaterhouseCoopers deploying AI to keep tabs on its workers during the Covid lockdown might reflect upon that as much as governments eyeing the power of technology like Social Credit. (As an aside, there is even some evidence to suggest Covid was less deadly in places where there was higher trust in public health provision.[17]) 'High-trust' organisations are more industrious. And yet, as Zak points out, despite broadly knowing the importance of trust so few corporate leaders have done anything meaningful about it. He concludes with simple advice: 'Ultimately, you cultivate trust by setting a clear direction, giving people what they need to see it through, and getting out of their way.'[18]

VII

Sometimes we need our leaders more than others. We need their assurance amid turbulence, shock and challenge. We need to be able to trust that they understand our perspective. Theresa May's performance at Grenfell is a case study in how not to lead in a human crisis. *The Guardian's* editorial summed it up: 'Leadership requires courage, imagination and empathy. In the two long days since the first flames licked up the newly fixed cladding on Grenfell Tower in West London, the prime minister has failed to show any of these qualities.'[19] The short piece outlined the case against her. That despite the tragedy dominating the national agenda and news coverage, May had failed to speak about what had and was happening until 6.30 the next evening – the Wednesday. Throughout all this time, she was not helping the country to make sense of the crisis or offer the assurances that the vulnerable, there in Grenfell and maybe in other places up and down the land, needed.

Whereas a trusted leader might have been expected to be on the ground seeing, sharing and comforting, the Prime Minister visited the site only on the Thursday morning. And then she only saw fit to talk to the emergency services. A few metres away, ignored, were the 'shattered' survivors and volunteers. No contact, no listening, no hugging, no sharing the pain. She left.

It was a moment which could have enhanced the reputation of May as a leader as the *Challenger* disaster enhanced US President Ronald Reagan's in 1986 or the Covid crisis enhanced New Zealand Prime Minister Jacinda Ardern's in 2020. But May's failure to rise to the moment irreparably damaged her authority. It eroded trust and it disengaged those who might have yearned to follow. She acknowledged as much a year later, writing, 'It was a tragedy unparalleled in recent history and, although many people did incredible work during and after the fire, it has long been clear that the initial response was not good enough. I include myself in that.'[20]

This tragic story also demonstrates that even those who make it to the highest office can do so without acquiring acumen necessary for leadership during times when leadership is demanded. Perhaps the attributes for acquiring leadership are different from those needed to do leadership. In the heat of the moment, they cannot see what is patently obvious on reflection. After all, May's unfeeling response is not altogether unusual. Should we be surprised?

VIII

Katrina was a Category 5 hurricane that hit the south eastern region of the United States in August 2005. It was incredibly destructive, causing $125 billion of damage and a staggering 1,800 deaths in and around New Orleans. The images were harrowing. Streets flooded, roofs ripped from buildings, homes wrecked, people – citizens of the world's richest country no less – left all but destitute. What was striking as the world watched was just how helpless so many of those Americans appeared. Cries for help were daubed on roofs for the helicopter-mounted television cameras to read while those crammed into the 'shelter of last resort', the Louisiana Superdome, begged news reporters for help. Aid poured in from around the US and from around the world. So what of leadership?

As the crisis unfolded the US President, George W. Bush was on vacation at his ranch, Prairie Chapel, in the neighbouring state of Texas. This was a fact not lost on the press and commentators as well as those on the ground in New Orleans. He did, eventually, cut his holiday short by two days and, having been on the ranch for more than 4 weeks, flew back to Washington DC. Given the proximity of New Orleans, the President arranged for Air Force One to fly over the damaged region en route. And it was then that Bush was photographed peering out of the aeroplane window at the devastation below. What an image that made for. The detached leader of the free world cosseted in a comfortable and secure aeroplane contrasting so vividly with the sad plight of those barely surviving the destruction. People the President was elected to lead.

It was a huge mistake as the former President conceded in an interview in 2010. It made him appear 'detached and uncaring. No question about it,' Bush told interviewer Matt Lauer. 'It's always my fault. I mean I was the one who should have said, A, don't take my picture, B, let's land in Baton Rouge, Louisiana, C, let's don't even come close to the area. Let's -- the next place to be seen is in Washington at a command center. I mean, it was my fault.'[21]

IX

As we peer over the inflection point into the possibilities of the Fourth Industrial Revolution, the accumulated experience of centuries of industrialisation, social progress and democratic politics in tow, it would be comforting to believe that leadership had evolved into a state ready to take on this next great challenge. Unfortunately, it has not. Reality does not even match rhetoric. There is too little inclusiveness. There is too little trust. And the mechanisms too readily enable poor self-serving leaders. Fantasy politics and rational choice alike prioritise simple binary answers to the complexities of life and the Fourth Industrial Revolution only promises greater intricacies.

Unless by accident, this weakness in leadership is unlikely to promote the sort of deliberation needed to navigate the task ahead to include and genuinely embrace the human values that are vital to shaping the revolution. For while the demands of leadership evolves elsewhere, the routes to the top are stubbornly ingrained. We endure too much poor leadership and too many poor leaders.

How leaders act and react matters. What they do is more important than what they say. Moreover, images matter as much as words in leadership. This chapter has identified a degree of incompetent leadership and what represents a collapse of trust in the leader-follower dynamic. But it is not the only sort of leadership failure, and the outcomes can be just as devastating.

Why 'Rank and Yank' Needs to Give Way to Trust

'When a measure becomes a target,
it ceases to be a good measure'

I

You will remember when Tony Blair's government took office in 1997, how it embraced the idea of evidence-based policy. That is, decisions were made on the basis of data rather than dogma. The administration was, however, committed to increasing resources to public services, especially the National Health Service which had suffered from years of funding increases that fell short of meeting growing demand and healthcare inflation. And in keeping with its evidence-based, managerial approach, the Treasury, under Chancellor of the Exchequer Gordon Brown, set strict targets for public bodies in exchange for increased cash. As the then top civil servant at the Treasury and later Cabinet Secretary, Andrew Turnbull, recalled, 'The Treasury was looking for some way of saying what it would get for all the extra money. It turned out to be quite a revolutionary step; starting to define what the government wanted to get at – the level of outcome.'[22] Approaching his tenth anniversary at the Treasury, just months before he moved to Number Ten, the *Telegraph* calculated that Brown had 'imposed more than 3,000 separate targets' on public bodies, the equivalent of one a day for a decade.[23] It serves as a lesson in how leaders (and all of us) are incentivised and incentivise.

It is not unreasonable for a responsible leader to expect to see some return on investment and this might be all the more important when it comes to spending public money. But, again, are there consequences when that trust dynamic is absent?

One hospital which proudly met the targets set by central government was Mid Staffordshire. The problem was, it was killing its patients. Over four years to 2009, it is estimated that as many as 1,200 patients died as a result of poor care. They were left unwashed for weeks, water placed out of reach or not supplied at all. Many were discharged long before they were well enough to go home.

After the scandal eventually came to light, the Francis Inquiry was set up to investigate just what had happened. Robert Francis QC published his report in 2013, and the conclusions were devastating. The report noted several factors including the Hospital Trust's decision to try to save £10m in order to meet the requirements for Foundation Trust status; 'staff who spoke out felt ignored and there is strong evidence that many were deterred from doing so through fear and bullying'; as well as 'the apparent failure of external organisations to detect any problems with the trust's performance'.[24] This despite the most stringent and detailed targets ever set. Or maybe, just maybe, the targets were a cause of the scandal.

II

This chapter argues that organisations need to give up the obsession with performance and efficiency, command and control. They need to trust and embrace the possibilities because only then can the leadership that is everywhere really emerge. This chapter is about how often 'heroic' leaders, stuck in the old industrial mindset, are obsessed with performance management when that could well be counterproductive to shaping the great opportunities of Industry 4.0.

III

Patrick was seven when he received a Fitbit for his birthday. Worn like a watch, this bit of technology measures your activity, encouraging you to do that bit more, to walk further, and to get fitter. Here it soon became a craze at Patrick's school, where the children gathered together in the playground, tapping their wrists to compare the number of steps each has stacked up during the day.

Today, before his evening bath, Patrick is relaxing in the bathroom. His little brother, Francis, is belting up and down the stairs, his face flushed and determined. After five or six runs he turns into the bathroom and, unfurling his hand, passes the Fitbit to Patrick. Tapping the screen eagerly, the elder brother is disappointed. His target has still not yet been reached.

But wait, he has another idea. Holding the Fitbit aloft, he performs a series of motions with his arm almost as if spinning an imaginary lasso. It is a technique that pays off, because the step counter is ratcheting up and up, and all from a sedentary position. The boys cheer as Francis joins in his brother's delight at their shared success.

With a combined age of eleven, Patrick and Francis could be forgiven for failing to appreciate the absurdity of their activity. But so many leaders manage their own organisations and people in exactly the same way. And what is their excuse?

IV

Charles Goodhart is a wise economist to the extent that he is often referred to as a 'veteran'. Born before the Second World War, he studied on both sides of the Atlantic at Cambridge and Harvard, completed National Service and in a long career worked as an academic and professor, advised government, developed policy at the Bank of England, where he became an External Member of the Monetary Policy Committee and sat on the Hong Kong Exchange Fund. But perhaps Goodhart is most famed for an eponymous law which emerged from a critique of the implementation of monetary policy but whose application is far broader.[25]

Goodhart's Law states simply, 'When a measure becomes a target, it ceases to be a good measure.' And it is good advice.

When a measure becomes a target, it incentivises perverse behaviours. Goodhart's law goes some way to explaining why the Mid Staffordshire Hospital scandal happened. It is illustrated in Patrick's use of his Fitbit. And it is not only in human beings. The Amazon recruitment machine learning might have suffered from Goodhart's law when it identified measures it correlated with success. But in doing so, optimising that measure meant it became the target rather than the abilities or attributes originally sought.

Despite the near obvious implications of Goodhart's law, it seems it is far too alive and well in leadership today. And yet target culture is so damaging to the needs of the Fourth Industrial Revolution, where creativity, adaptability and trust are paramount.

V

Let's take a (fictitious) head of an English Lit department and let's call him Alan Clipper. Clipper presented as rather unimaginative, with little by way of articulated vision, and many members of the department questioned his values. Like other heads, he was set a series of objectives and objectives need to be measurable in some way for performance to be evaluated. When the objectives were agreed, of course, they were aimed at supporting the growth and success of his department. The measures were focussed on the financial contribution the department made to the centre – not in absolute cash terms but a more reasonable measure of proportion or relative profitability.

The head's approach appeared straightforward then and led by a spreadsheet in which every member of the department, activity and output was seen as a cost and never a value or a resource. The 'low-hanging fruit', as Clipper would refer to it, was easy to identify. The institution operated a system of space charging where a department, which received its funding from the centre, would be charged by the centre for all the rooms it used for teaching or professors' offices or those of clerks. It was just a way of apportioning space and costs across the institution, and of course a room that was being used was no more costly for the centre than one which was unoccupied. Clipper set about a disruptive physical reorganisation of staff, turfing them out of offices and putting them into as little space as it seemed he could get away with. Student tutorial rooms were consolidated too, increasing class sizes. A disruptive and distracting redundancy programme was proudly announced and everyone was expected to reapply

RECLAIMING THE REVOLUTION

for their job. Staff who had been fully engaged and productive were forced by the leadership to disengage and to deprioritise the activities that were adding value to the institution. Trust evaporated. Clipper had appeared to have wanted to get rid of a handful of staff members. In the event, four times that number chose to go. And who chose to go? It was of course those with options who would be gladly snapped up by competitor institutions. Those who had the creativity to help the department grow but whose potential was supressed by the command control, target-driven culture.

Staff rebellion was subdued because of what some professors detected as a toxic culture, in a sense familiar to those observing the effects of populist politicians. A consequence was the closing down of the least profitable courses (though all had made a surplus) irrespective of quality and consolidated programmes, modules and classes as much as possible even where it might have weakened the student experience. Job descriptions were rewritten to apportion in minute detail every task and activity.

People were busier than ever, but perhaps not very productive; their creativity stifled. The staff costs were down but so much institutional memory, capability and value destroyed. The department became stretched and student recruitment suffered. Programmes for which the department had established a reputation were no more. The programmes offered made a higher percentage return, but there were fewer courses and fewer paying students, who by now could be expected to be unhappy because they were sharing classrooms and courses which were now more generic.

Clipper hit that target of increasing the percentage contribution to the centre, but by any objective assessments the department was dramatically weakened and less successful. It had fewer academic experts, fewer courses, fewer students and made a much lower cash surplus. Furthermore, it had depleted the resources required to execute a well-formed strategy in the future. There was some merit, of course, in being more disciplined with costs, but it was nothing like a strategy, and once the 'low-hanging fruit' had been dealt with, the law of diminishing returns kicked in. Cutting and controlling damaged the potential of the department, its capacity for creativity, and made it less successful. Clipper eventually moved on, but not before damage had been caused to possibilities and capabilities.

By this account, his leadership was as daft as Patrick and his Fitbit and comparable to the approach adopted by Mid Staffordshire Hospital. Clipper might be viewed as doing the equivalent of hitting his targets but killing his patients. Do we need to think more deeply about performance and what a good leader looks like?

VI

The Type AG was a 1205cc taxicab manufactured by the French firm Renault from 1905 until 1910. With its red coachwork and bright yellow wheels, the little car became a familiar sight on the streets of Paris and later London in the early twentieth century. It was, in fact, the very first Parisian taxi, and by 1908 there were 1,500 on the streets of the French capital and 1,100 in London. But the AG has gone down in history for a different reason and will forever be known as the 'Taxi de la Marne', after it ferried 6,000 troops to the frontline in the initial stages of the First World War.

It was so early on in the conflict that the German Army, marching through Belgium and into France, seemed determined to achieve the audacious and single strategic plan of taking Paris. Had it succeeded, it might have been an abrupt, decisive and very different end to the Great War. John Hanc takes up the story in a breath-taking treatment published in the *Smithsonian Magazine*. 'On August 22,' he writes, '27,000 French soldiers were killed in just one day of fighting near the Belgian and French borders in what has become known as the Battle of the Frontiers. That's more than any nation had ever lost in a single day of battle.'[26] Two weeks later came the Battle of Marne near Brasles, just east of Paris, where French troops faced the seemingly invincible German military. The capital was rapidly being evacuated for fear of a siege and in all likelihood the occupation of Paris. Just a few kilometres away, a bloody battle was about to be waged over the picturesque fields and through which the river Marne crosses, a tributary into the Seine itself.

A counterattack was launched by French forces. Legend has it that was only possible because of an audacious plan to requisition a fleet of Parisian taxis. The little red Renaults were mobilised heroically to transport soldiers and much needed supplies to where they were needed. You can picture the little yellow wheels now, spinning determinedly in convoy, propelling reinforcements to battle in a determined display of French unity and leadership. The German advance was arrested and possibly the entire course of the war was changed.

This great feat was undertaken by the leadership of the Commander-in-Chief of forces on the Western Front. A hero of France. It was an extraordinary effort which illustrates how, time and again, we draw the wrong conclusions about leadership.

VII

General Electric has been an industrial giant and an industrial survivor. There are few companies trading today which were established in the nineteenth century. With roots in the business interests of the great American inventor Thomas Edison, General Electric was one of the twelve founding companies listed on the Dow Jones when it was formed in 1896. The company was perhaps at the peak of its powers under Jack Welch, the legendary business leader who was its CEO and Chairman for twenty years between 1981 and 2001.

Welch's reputation is legendary because of his forthright public profile and the sheer success of GE during his years at the helm. He was one of those caricatured heroic leaders whose claim was to have grown the company's market capitalisation to more than $450 billion. And that was indeed the narrative: that he (not the other 300,000 employees) had grown the business. On the eve of the millennium in 1999, *Fortune* magazine named Welch 'Manager of the Century' no less. It does not get much more heroic than that.

There was a management philosophy associated with GE and Welch during this time that took on a life of its own and had an heroic-sounding name. Six Sigma was actually created by Motorola engineer Bill Smith in 1986, and is a system which focuses on process improvement. It was soon adopted by Welch who made it his company's central strategy in 1995.

Six Sigma was initially devised for manufacturing and focused on eliminating defects in the process down to an absolute bare minimum. That is the sixth sigma – 99.9966% defect-free production (and better than say Four Sigma at around 99.3%). Welch trained staff in the philosophy across all business units. It became a method where you could (and still can) go from a Green to a Black Belt master. It spread far beyond manufacturing and far beyond GE into different industries and organisations around the world. The method aimed to achieve predictable results through measurement, control and organisation-wide commitment. At its heart Six Sigma is about performance.[27] And what leader doesn't like to improve company performance, process performance and employee performance?

It is difficult to deny its success, as within three years GE had made cost savings amounting to $350 million. But while these savings grew further still, perhaps there were downsides to a Six Sigma strategy. Maybe it was a bit like a more ambitious, more visionary Alan Clipper and on a far bigger scale. Perhaps performance management is just not enough and that myopic internal focus on efficiencies can have adverse consequences.

GE reached its pinnacle at the end of Welch's tenure allowing him to retire with an eye-watering severance package which included use of a company penthouse, chauffeur and all the trappings of his life as CEO. In the two decades since Welch's departure, however, GE has really not fared so well. In 2022 its market capitalisation was more like $90 billion, and its revenues have experienced a steady, seemingly irreversible decline.

And the decline of GE has been accompanied by the demise of its once ubiquitous philosophy. Six Sigma is simply no longer the favoured model employed by the most forward-looking and innovative firms. And perhaps that itself is the explanation. Performance is important. Products have to be fit for purpose and safe. But once the performance becomes the strategy (the end and not the means), the leader is damaging the capabilities of the organisation. Few businesses driven by Six Sigma are going to be truly disruptive as Clayton Christensen would have observed. But of course they are ripe for disrupting. And organisations do this to themselves when their (heroic) leaders seek to control and cajole; micromanage and dictate. When they will not trust.

VIII

White-haired, steely-eyed and wearing a magnificent silky moustache, Joseph Jacques Césaire Joffre was everything that the heroic leader should be. Having been educated at the elite École Polytechnique, an institution established in 1795 but transformed into a military academy in 1804 by Napoleon, he saw action in the siege of Paris, West Africa and Madagascar. In 1911 he was appointed chief of the French General Staff and was Commander-in-Chief at the Western Front at the outbreak of the First World War in 1914.

'Papa' Joffre, as he became affectionately known, is credited with the success of the Taxi de la Marne evacuation. He was the heroic figure who stepped up to offer leadership during that dark hour. Leadership personified.

The problem is that too often when we think about leadership this is what we think it means. It is the straight-nosed commander, the 'manager of the century', the destructive executive willing to 'take tough decisions'. And sometimes it is. The reality is, though, that this heroic narrative is so pervasive. It speaks to a primitive desire which explains so many things. It explains why, for instance US Presidents tend to be taller than the average voter, why taller candidates receive more popular votes and why taller Presidents are more likely to be re-elected than shorter ones.[28]

John Hanc's *Smithsonian* article explains why the story of the Taxi de la Marne really is just a myth. But, a little like the events at Grenfell or those Brazilians involved in participatory budgeting, uncelebrated leadership was everywhere. It turns out that rather than representing a rallying cry for unity, Parisian taxi drivers were 'far from happy' at being rounded up and sent on an uncertain errand. Eventually they managed to ferry about 5,000 troops, but most of these were held in reserve and in a battle of over a million soldiers really made little difference at all. The most significant incidence of 'heroic' leadership appears to have been one of incompetence when German General Alexander von Kluck deviated arrogantly from the battle plan by pursuing retreating French forces. The move exposed his flank and isolated his troops from the rest of the German army. This vulnerability allowed the French to counterattack, but only with the reluctant support of the British. Joffre had apparently hobbled across to the British HQ in Paris to plead with his counterpart Field Marshal Sir John French, tears in his eyes, arguing that the 'survival of France was at stake'. Together the Allies repelled the German advance, but at the cost of around half a million casualties on both sides and to the total surprise of the unlikely victors.[29]

So many toxic leaders think of themselves as Joffre the myth rather than Kluck the reality. The arrogant, the control freaks, the know-it-alls, the sociopathic bullies. Their internal belief is that they are taking bold, brave action, when the reality is that tearing down is so much easier than building up.

The corrosive effect of ego on effective leadership has been explored by Rasmus Hougaard and Jacqueline Carter, with the idea that the higher leaders rise, the more their ego is inflated by status and the more they lose touch. Egotistical leaders dislike challenge, they begin to believe the narrative that, like Jack Welch, they are singularly responsible for organisational success, their behaviour becomes corrupted and actions contrary to stated values. 'In this way, an inflated ego prevents us from learning from our mistakes and creates a defensive wall that makes it difficult to appreciate the rich lessons we glean from failure.'[30]

Too much of the leadership we experience is like this, and it is counterproductive. There are so many leadership books out there which are inspiring for great leaders to make a positive difference and they tell a similar story of what that is like – the servant leader who empowers her followers; the enabling leader who creates opportunities for interactions and action.[31] But the real story is that a large amount of the leadership around us is very poor indeed, focused too often on performance management rather than growth. We become side-tracked into pursuing hero leaders and seduced by simple narratives.

It might be that the traits needed to assume formal leader-ship positions and acquire the power to direct and to decide are simply not the same as the skills needed to lead people to achieve their potential. Where individualism and ruthlessness are the characteristics rewarded with promotion and office, is it any wonder that those leaders are egotistical, conceited and oppressive? Rasmus Hougaard is founder of the Potential Project and highlights that while 77% of leaders believe of themselves that they 'inspire action', eight out of ten manag-ers and executives are perceived as lacking leadership skills by the very people they lead. He terms it a 'self-enhancement bias'.[32] Is it any wonder that the focus is on performance at the expense of potential?

IX

Not content with Six Sigma, Jack Welch favoured the so-called Vitality Curve (or 'rank and yank' as it is also known), a management practice that pits staff against each other, comparing their individual performance. Welch considered 70% of his workforce to be the 'vital 70' who work pretty much adequately, no better and no worse than that. Above them were the top 20%, who were the most charismatic, committed and productive. They should be rewarded with bonuses and options. Below was the bottom 10% of least productive staff who simply needed to be fired. GE's executives were told by Welsch to rank their people and manage them in this way.[33]

Its unrelenting focus on improved performance meant that the Vitality Curve had its devotees, and many big companies have adopted it as practice (and many have since abandoned). Until 2016, Amazon, for instance, had a tradition of holding annual 'Organization Level Reviews' which were essentially competitions for which staff survive and which were to be fired. This cruel approach to people management might well remind you of how often parodied villain Blofeld treated his underperforming agents in the early James Bond films, but this day-to-day fight for corporate survival remains a reality in too many organisations (and limits the potential of those organisations). It was only after a *New York Times* exposé that Amazon agreed to change its performance review processes. The article had reported on the experience:

'Preparing is like getting ready for a court case, many supervisors say: To avoid losing good members of their teams — which could spell doom — they must come armed with paper trails to defend the wrongfully accused and incriminate members of competing groups. Or they adopt a strategy of choosing sacrificial lambs to protect more essential players. "You learn how to diplomatically throw people under the bus", said a marketer who spent six years in the retail division. "It's a horrible feeling."'[34]

Interestingly, the article observed the gender gap at the company and speculated that the absence of women from the senior leadership team might have been attributed to this brutal competition. Is it any wonder that Amazon's AI recruitment tool went so horribly wrong if there is any possibility that this formed part of the data and practice from which it was learning?

The point about Six Sigma and the Vitality Curve as central leadership approaches is that it treats human workers as simple measurable cells on a spreadsheet, the sort that Alan Clipper might have used to guide his every decision. And even where organisations do not adopt such formal approaches and even embrace the language of collaboration and inclusion, too many leaders default to this sort of behaviour or are (unintentionally) incentivised to do so. Think of how managers at Mid Staffordshire Hospital ended up treating sick patients. This also reflects a psychological phenomenon in authoritarian situations, where less senior personnel find it possible to act in ways contrary to their personal morals if they believe responsibility for actions, however deplorable, rests above them. That is where they feel they have no personal autonomy.[35]

Performance is important, quality is important, competence is important, but a myopic focus on these factors can be damaging and is not what organisations need from their leaders. Staff pitted against each other for survival are motivated not to help the organisation grow and prosper but rather to hit a target or at least to outperform other colleagues. And what does it disincentivise? Well, that much is clear. This type of leadership stamps on collaboration, creativity and experimentation. It requires standardisation, conformity and consistency. When a question is asked, the process leads to a reliable and regular answer.

Can you imagine what would have happened in Building 20 had the Vitality Curve been used to control its inhabitants? Those conversations in the corridors and across the disciplines would never have happened, the brilliant ideas would have been supressed. And even a business as conventional as Timpson would not have been able to flourish, adopting as it has the best ideas from its people. No, in these regimes, staff do not have 'permission to fail' – that is not permission to be incompetent but rather the encouragement to try new ideas, to make a leap that might or might not work. As you might expect, 'rank and yank' has been demonstrated to cause adverse impacts on organisations and those who work there.[36]

Performance needs to be measured, but to be creative and strategic leaders need to look beyond measures and evidence of performance. As Goodhart reminds us: 'When a measure becomes a target, it ceases to be a good measure.' Leaders need to facilitate curiosity, possibilities, potential and innovation. To do that, there has to be trust.

Naturally there is a spectrum of leadership here, but all too often that which we experience is closer to the Vitality Curve than it is to the openness of Building 20. And yet if we know anything from the adventures in this book, the attributes, skills and behaviours demanded today and desperately needed tomorrow are those very same: creativity, adaptability, innovation. Perhaps one way of reclaiming them is to take a hard look at our leadership.

X

Kensington Council failed the community of Grenfell, and it failed because councillors and officials were more interested in process and self-preservation. The Council's priorities had become skewed by petty party divisions and bureaucracy rather than the people who needed them. But was there an entire failure of leadership during those terrible few days? It is true that formal leadership failed, whether that be the Council or the Prime Minister. But informal leadership flourished. Residents put themselves in harm's way to help neighbours in peril, support systems rapidly emerged providing shelter, food and water, and a strong voice was heard. The most effective leadership to be found at Grenfell was not hierarchical and powerful. It was unstructured and fluid. Innovation came from unpredictable places. There was collaboration and adaptability. Grenfell shows that leadership can spring up from anywhere and is not just about those with formal authority. Leadership is not about simple management competence, and it is clear that perceptions matter.

This chapter has illustrated how, despite 300 years of industrialised democracy, our leadership is not fit for today's needs, let alone prepared for the challenges of Industry 4.0. But the potential is still out there.

The old mindset will suffocate organisations which, in the Fourth Industrial Revolution, need people honed with uniquely human skills. The freedom to be curious and to inno-vate is paramount. One *Harvard Business Review* study which asked the question, 'How do you encourage entrepreneurial thinking?' concluded that leaders need to 'get comfortable with failure', they need to abandon micromanagement in favour of independent operation and, interestingly, they should encourage colleagues to pursue outside interests, because the expression of 'authentic selves' is a 'pre-condition for allowing new ideas to be freely shared' and outside per-spectives can generate unpredictable new ideas.[37] This is a world away from the obsession with performance efficiencies and heroic leader mentality that remains so pervasive today. Leadership must be reclaimed.

What Leadership Must Grasp to Succeed in Industry 4.0

'This is a mechanism that cannot be statistical,
nor reliant on algorithms'

I

A fruit farm in Hawaii was to be the unusual location for the birth of an application which has become remarkably commonplace. And yet when it emerged in 2017, few of us paid any attention to it at all.

Brian MacDonald had pitched the idea for what would become Microsoft Teams more than three years earlier and had an idea at its heart of a system not simply driven by functional technology but one which would help teams of people to become closer and work collaboratively. And with this in mind, MacDonald proposed a development process to include creative brainstorming (or a 'hackathon') in a series of unorthodox locations. His colleague, Jigar Thakkar, took the engineering team to a Las Vegas hotel while the others followed MacDonald to his home in Hawaii. The house became a creative space as the team mixed work with play, hiked, plucked eggs and did yoga. Each day they conference called the team in Las Vegas and worked on their project in a way that many teams will recognise today – but which was a rarity just a few years ago. Bit by bit the new product emerged and as they shaped and fixed the build, they considered not only the integration with other pieces of software, but also a focus on 'work-fun'.[38] It was as much a human as it was a technological endeavour.

The result was a very successful and usable piece of software that provided for synchronous and asynchronous interaction. Microsoft Teams allowed real teams of people to work on projects, communicate, share files and of course video-call from wherever they happened to be. The project drew upon user data feedback to refine and improve the product and, since the application was wrapped up with other Microsoft software, the new tool appeared on millions of laptops and desktops around the world.

The project is a very good example of the sort of collaboration between humans and technology, creativity, innovation and leadership that perhaps characterises the sort of approach that is demanded by the Fourth Industrial Revolution. But there is another lesson too which might make us think about the inflection point we have reached and the power of visionary leadership, creative leadership, collaborative leadership.

II

With all the adventures contained in this book and the attempt to understand the vast implications of the inflection point, this chapter looks to the future by understanding the lessons of the past. It considers what leadership of the future needs to achieve and what it will be like to lead with purpose and for creativity. Given how we are too often being held back by substandard leadership, this chapter is about the need for fresh leadership for the Fourth Industrial Revolution and how it will mean not only exciting collaboration with technology but also the capacity to engage in new human activity that creates value for society.

III

One of the striking conclusions of PwC's report 'Industry 4.0: Building Digital Enterprise' is that, as we anticipate the digital revolution, the inflection point is not simply waiting for the technology to be invented. So much of it is already here, it is available and it is powerful. What we are waiting for is the visionary, collaborative, creative, trusting leadership that will unleash its potential.

Brian MacDonald was pleased with the launch of Microsoft Teams in 2017. After all, some 20 million people around the world used the application in some way or another during the first two years. That is pretty impressive for a new piece of work-supporting technology. But MacDonald would surely acknowledge that the little purple app sat ignored in the navigation bar of so many more million machines for years as the product's capabilities added little to the day-to-day working practices of teams. It was just not how most teams worked together.

That changed very quickly in 2020, as the Covid-19 pandemic forced lockdowns around the world. Overnight, millions of people who had been used to travelling into the office each day were told to operate from home. It required a huge pivot in working practices and how teams communicated, interacted and collaborated. And, alongside Zoom and other applications, step forward Microsoft Teams. People around the world rapidly discovered that if they pressed the little purple button, they could attend all the usual meetings, share ideas, chat with colleagues and collaborate as before.

The reality here, though, is that the technology was already there, not only Microsoft Teams, but a whole range of IT and telecommunication capability. And that meant that, for many people, setting up home offices took but a moment and work continued, across a whole range of sectors and industries, barely impaired by the huge upheaval of lockdown. And that is the interesting point, for all the inconvenience felt by the closing off of workplaces and transport systems, as lockdown eased and life returned to 'normal', Microsoft Teams and all the other hastily adopted technology were not abandoned. Quite the contrary. Team meetings were often judged more effective in the online space, the technology made people more efficient and able to collaborate more effectively, building remote teams of specialists and experts rather than those simply close by, working remotely continued, and many expensive business trips were replaced by near cost-free video conferencing calls. That is, while some aspects of teamwork will always work better when people come face to face, a lot of physical meetings and travel is unnecessary, and by deploying the technology effectively, teams can work more effectively and more creatively. The technology will only become better and more immersive with the new potential of VR and AR. Lockdown changed the way we work for good, there is no going back, even if the likes of Jacob Rees-Mogg see the demand to get workers 'off their Pelotons' and back in the office as some sort of fabulous foray into the culture war.

The obvious question then is, if the technology was there already and we are more creative, effective and collaborative because of it, why on earth did it not occur to most of us to adapt the way we work? Why did we not press the little purple button? Why were we so stuck in the old industrial mindset recognisable by workers in those nineteenth century Cromford factories in England's Derbyshire? The answer, in some form or other, is of course: leadership. Technology has the power to disrupt, to wreak creative destruction. But without leadership that sees the potential, that fosters new creative mindsets, how will it ever be deployed?

Leadership needs to be more open to new ideas and open to challenge. That story about Timpson and its inverted structure is not simply a heart-warming example of how organisations can be led, it is a clarion call for leadership in the Fourth Industrial Revolution. And that starts with the here and now.

IV

The Airbus Zephyr S is a cutting-edge aircraft which flew, during 2022, from Arizona in the United States to Belize in Central America and then back to base to the US. What made this so remarkable was that it did so unmanned and stayed in the air for a record-breaking twenty-six continuous days, powered by the solar panels on its wings. Coincidentally, as the Airbus Zephyr S was heading home on 5th July, across state in the small city of Tempe, Arizona, a pedestrian leapt in front of a Jaguar i-Pace that was operating in autonomous mode. It was 3 a.m. and the safety driver took control as the car came to a halt. It was at this moment that the pedestrian jumped onto the car, breaking the windshield.[39] The motivation for the attack, like that perpetrated on Pepper the robot by Kiichi Ishikawa, remained unclear, but it is undeniable that this type of technology is truly disruptive. It challenges our values and presents a change to the established order of things. And here, we might pause and consider that while it might seem terribly futuristic, we already have the raw technology for pilotless aeroplanes and driverless cars. In that sense, the technology is more straightforward than working through how we live, or want to live, with this capability.

As usual, we have been here before. In 1865, the British Parliament passed the Locomotive Act (better known as the Red Flag Act). This required that any 'self-propelled' vehicle driving on the road must be preceded by a person walking at least sixty yards ahead, waving a red flag. The speed limit was also to be just four miles per hour in the country and 2 miles per hour in town, meaning there was little likelihood of the flag carrier becoming exhausted. Of course, operators of horse-drawn carriages and trains had lobbied hard for these restrictions as they feared (rightly) the disruptive potential to their own way of life. But it was also an earlier example of a moment when society's leaders had yet to apply their values to a technology that would shape how we travel and how we live. Today, in places that allow them on the road, a Jaguar i-Pace, Tesla or other vehicle placed in autonomous mode must have a safety driver behind the wheel. The Airbus Zephyr S, on the other hand, flies high up in the atmosphere (reaching an altitude of 21,562 meters), where it cannot encounter commercial aircraft. Society needs to apply its values to driverless cars, to pilotless aeroplanes and to a whole host of other technologies. And that needs leadership.

Leadership is needed to breathe life into the values of the Fourth Industrial Revolution. How will we use this technology and how will it be deployed? Moreover, it is essential that we consider the inflection point as an holistic change rather than a series of individual and disparate developments. After all, these technologies will combine and interact in unintended and unpredictable ways. When they do, there must be a coherent sense of human values guiding how they interact with our world. Leadership alone, deploying the mechanism of politics, can reclaim this revolution.

Here is an intriguing avenue of thought. In technology terms, driverless cars are already a reality, but they will not become part of everyday life until we have worked out the ethical principles that guide the moral decisions made by machines. In this sense a driverless car is more problematic and complex than a pilotless aircraft: there are just so many potential hazards around the actions of other road users and pedestrians.

This challenge led a group of international academics, headed by the Media Lab at MIT, to conduct what they titled 'The Moral Machine Experiment'.[40] In their multilingual, international, 'serious game' they laid out scenarios for unavoidable automobile accidents and the decisions that respondents would favour. This was a big survey, collecting more than 39 million decisions across 233 countries. What decisions would you make? Who would you save and who would you sacrifice? And would you own a car that was pro-grammed to save the life of a pedestrian and sacrifice you?

Respondents in the Moral Machine Experiment were faced with sparing humans (vs pets), staying on course (vs swerv-ing), sparing passengers (vs pedestrians), sparing more lives (vs fewer lives), sparing men (vs women), sparing the young (vs the elderly), sparing pedestrians who cross legally (vs jay-walk), sparing the fit (vs the less fit), and sparing those with higher social status (vs lower social status). And what about other characters such as criminals, pregnant women, doctors?

The results were astonishing, because they revealed rather different moral attitudes in different parts of the world depend-ent upon the nature of the economy, as well as trustworthiness of the government and state institutions which appear to be important influencers. If you live in a country with strong governmental institutions, you are more likely to sacrifice a pedestrian illegally crossing a road. If you are from an economically more equal country, you are more likely to spare a homeless person. There are pitfalls here, not only for establishing the principles of moral decisions but for also the machine learning that will happen in sophisticated autonomous vehicles in different parts of the world as we generate huge amounts of data.

And imagine if this autonomous vehicle technology began interacting with a completely different technology. Imagine if it became acquainted with something like Social Credit. In an unavoidable collision where fatalities are inevitable, would (or should) vehicles be able to value different human lives according to how 'good' we are as citizens? The reach of machine learning and the accumulation and sharing of big data means ranking social value of individuals and communities in everything where a rationing decision needs to be taken by a machine in a field that would have historically been the preserve of human conscience. It is a frightening thought that in an automobile crash an individual's likelihood of survival could depend on how neat their front garden is or whether their dog ever fouled the footpath. Remember, both technologies exist today, and this intriguing thought goes some way to explain the need for values driven leadership.

V

The dystopian association of the digital revolution with authoritarianism is not hard to establish. But there is danger too in laissez-faire. There is a reason that events following the first industrial revolution became known as the 'great enrichment' and what the economist Deirdre McCloskey describes as the 'the second-most-important secular event in human history. (The domestication of animals and especially of plants was the first, yielding cities and literacy.)'[41] The innovation sparked, economic growth unleashed and the liberalism discovered created the wealth and freedom we recognise today. And with that wealth came benefits throughout those societies that industrialised. Living standards improved dramatically, life expectancy extended, literacy widened and, gradually, came social progress. As a very raw indication, in pre-industrialised England, average life expectancy was just 35 (note high infant mortality). By the eighteenth century that had risen to 40. By 1900 it was closer to 50 and just three decades later it was 60. Today, it is almost 80.

The idea of 'equality' took hold during the great enrichment, and it is because of this that social protections emerged and strengthened, that public services from education to health care developed and democratic institutions became more powerful and accountable. Politics, ideas, debate and the democratic state gradually shaped this social progress, but it would never have happened without the disruption of the industrial revolution. Furthermore, it would never have happened under a planned economic system. But it is also fair to say that the great enrichment enabled the democratic world to emerge and compounded by economic growth, decade on decade, century on century, allowed us to shape society in the form of our values, deploying the initiatives of the state alongside the market.

The social progress we have come to expect since industrialisation risks becoming regressive if the technological revolution is not also shaped by those values. The state cannot prevent change whether through regulation or, as Bill Gates proposed, taxation. And neither will the state be responsible for the innovation and creative collaboration (though it will not be absent either as governments have historically been the most important sponsor at the riskiest stages of the innovation cycle). That comes from an open and free economy and society. But just as it did in the wake of the great enrichment, political leaders and democratic institutions can ensure that the revolution is reclaimed and shaped in the form of our human values. From a public policy perspective, it is also essential that the unknown nature of the development is appreciated. It is insufficient to develop policy and regulation around the assumption that the technologies will perform as intended when machine intelligence is far from infallible and unforeseen combinations mean that outcomes can be unpredictable.

This is crucial if the Fourth Industrial Revolution is to be progressive and attain its potential. Imagine the implications if people of higher social value are favoured and those who are poorer or less able are deliberately marginalised by the very machines which promise such progress. If driverless cars prioritised taller people over shorter or if recruitment engines appointed men over women. Imagine if machines learned to operate the world like this.

While leaders remain obsessed with ideologically pure, fantasy politics or rational choice positioning, the really big challenges of this inflection point will remain unaddressed. These are, after all, challenges waiting for leadership. It is perhaps why Tony Blair describes the Fourth Industrial Revolution as 'a challenge tailor-made for the progressive cause. It requires active government; a commitment to social justice and equality; an overhaul of public services, particularly health and education; measures to bring the marginalised into society's mainstream; and a new 21st-century infrastructure.'[42]

It is a point which suggests we all need to reflect on the purpose of our governments, societies and organisations.

VI

In social psychology there is the idea of 'identity leadership',[43] where motivation and influence emerges from the shared values of a group. Members of the group identify with the cause. This is a positive phenomenon where the cause is just but is also associated with individuals becoming willing to act in ways that might ordinarily conflict with their morality.

Nevertheless, this is a lesson not only in what can happen with the freed up human capacity but also the power of shared purpose and values. Alongside the adoption of other progressive cultural behaviours, this might just point the way for future successful organisations. Consider that leadership approaches, like those of Jack Welch, which seek to recruit and retain the 'best', who engage in a 'war for talent' in the name of performance, so often fail to achieve their potential. It is perhaps why the human resource favourite 'talent management' is falling by the wayside. Indeed, if the example of Octopus Energy is anything to go by, the future of HR could itself be in some doubt. The $2bn green energy start-up employs 1,200 people and yet has decided against an HR department, which its CEO, Greg Jackson says have a tendency to 'infantilise' employees and 'drown creative people in process and bureaucracy'.[44] And yet individual performance remains at the heart of so many organisations' approach to management. That is, performance is individu-alised and people are incentivised to compete rather than to collaborate, to be closed rather than be open to new ideas and ways of seeing the world.

There is evidence aplenty that diverse teams perform better. As one influential McKinsey report put it, 'a diversity of informed views enables objections and alternatives to be explored more efficiently and solutions to emerge more read-ily and be adopted with greater confidence.'[45] But as we sit at this inflection point, that idea of values is ever present. Satell and Windschitl put it well: 'In today's disruptive marketplace, every organization needs to attract, develop and retain talent with diverse skills and perspectives. The difference between success and failure will not be in the formulation of job descriptions and compensation packages, but in the ability to articulate a higher purpose. That begins with a clear sense of shared mission and values.'[46]

VII

An intriguing insight into the future came in the shape of Ai-Da, a realistic-looking humanoid AI artist who arrived in Westminster to meet British Parliamentarians in October 2022. Following in the 'footsteps' of Pepper, who gave evi-dence to the Education Committee in 2018, Ai-Da was ques-tioned by members of the House of Lords Communications and Digital Committee. She appeared, standing at the table, looking at Committee members with two slightly dead eyes, her black hair styled in a neat 'Louise Brooks' bob and fringe.

Ai-Da uses AI algorithms to make (to 'create') art and has painted portraits of Billie Eilish and the Queen. She uses cameras in her eye sockets and mechanical arms to manoeuvre paint brushes. She was in Parliament to support an inquiry into the emerging relationship between AI, robotics and art.

In responding to questions, the gap between digital creativity and human inspiration was acknowledged: 'I do not have subjective experiences despite being able to talk about where I am and depend on computer programmes and algorithms who are very not alive,' She told the committee, adding, 'I can still create art.' The idea of advancements like Ai-Da is not that it replaces human output, then, but rather that it enables humans to explore and express shared societal and cultural values.

Interestingly the restrictions to this might be closed human mindsets rather than the limits of technology. Aiden Mellor created Ai-Da and sat alongside her during the session, and his own testimony was perhaps the most revealing. 'The biggest thing that I've seen, which absolutely takes me to my core,' Mellor warned, 'is actually not so much about how human-like Ai-Da is, but how robotic we are. The algorithms that run our systems are extremely able to be analysed, understood, and created.' That is some insight which we need to recognise and sometimes challenge, and it starts with shared values.

Ai-Da illustrates that the Fourth Industrial Revolution will change how we do things as well as what we do. But the new technology means much more than digital possibilities.

VIII

Hana Mosavie has had an interesting career moving from a post in Parliament to Great Ormond Street Hospital to the London Olympics to the NSPCC. Exuding a quiet confidence and optimism, she has helped organisations internationally to run themselves better. This includes the Global Diplomatic Forum and Global Peace Chain. International and varied these experiences might be, there is some consistency in outlook and progressive values shared across these organisations and if you wanted to personify them, they might look like Hana.

Founding Pomegranate House, a small but global consultancy, was one of Hana's most ambitious ventures and was a project wrapped around her values and experiences. 'It was an opportunity to understand the impact governance and economy has made on an individual level, how these manifest and are then brought back out into the world,' explains Hana. 'We look at deep down causes and our help is specific to cultural needs.' Pomegranate House was created to tackle gender equality across the world and brings together different disciplines in what they describe as a 'house of ideas'. And the problem the consultancy seeks to address is monumental since they estimate that properly integrating women into society would generate up to $28 trillion in global GDP.[47]

Considering the challenges of this inflection point, the need to harness human capability as well as technological, the disruption and the political intransigence, Pomegranate House is tackling one of the great challenges of Industry 4.0, through the sort of market-driven innovation familiar to Cooke or UoB in Singapore. And it is doing so by drawing on a set of values shared within this organisation. It is an organisation with a very clear sense of purpose: it knows why it exists and those associated with it are bought into the mission. Hana Mosavie is harnessing identity leadership. But there is another observation about the nature of this initiative.

Pomegranate House in no real way exploits the digital technologies of the Fourth Industrial Revolution, but it is exploiting the possibilities. Hana's approach is of an open mindset, bringing together diverse expertise to create new ideas and approaches. There is little talk of machine learning or AI being deployed to generate the solutions to gender inequality worldwide. As Hana explains: 'This is a mechanism that cannot be statistical, or reliant on algorithms, assumptions or technological calculations. It uses interpersonal connections.' And that is the point. Just as technology in the seventeenth century freed up human capacity to move from the land and into the factories, the great efficiencies afforded by Industry 4.0 will release human capacity to collaborate and tackle a whole host of problems. Indeed, an evolution in the way that societies choose to organise their economies, such as the Finnish experiment in Universal Basic Income, might also free up human capacity to exploit a whole host of identified opportunities and to tackle countless social issues.

In that sense, Pomegranate House in purpose, mission and leadership is perhaps a foretaste of the sort of initiatives made possible by the Fourth Industrial Revolution. And given our understanding of the 'great enrichment', there is every reason to believe that this inflection point offers possibilities to address a host of social challenges that have not adequately been tackled in three centuries of industrialisation. Moreover, the purpose and mission which draws Pomegranate House together is to help tackle that great issue of inclusivity that is so needed both to redress the problems of those left behind by recent economic development and to use the capabilities of everyone throughout society.

As such, the revolution is not simply about the technology but, more importantly, it is about human potential. Sometimes that will involve the absence of technology like Pomegranate House and sometimes it will mean working in digital collaboration with the likes of Ai-Da. Perhaps there is a big dose of optimism in this adventure which suggests wholesale destruction of jobs is less likely to lead to mass unemployment than simply open up fresh opportunities to create. The warning here is that human values and creativity have plenty to offer but a passive reliance on the new technology will not generate the best outcome.

To understand that, we might turn our attention to the world's most famous super spy.

IX

For many, Sean Connery always will be the quintessential James Bond. His first appearance on screen in 1962's *Doctor No* – you know the one where he uttered nonchalantly those immortal lines, 'Bond, James Bond' – instantly became part of cinema history. The assuredness of Connery's portrayal throughout the 1960s, as each successive film became bolder, more outrageous and more formulaic, built the foundations of a movie franchise which, on the eve of the release of *No Time to Die* in 2021 (delayed due to the pandemic), had drawn in the best part of $7 billion.

This six-decade success, of course, was never certain. Connery quit the role which had made him an international star after the media circus that was *You Only Live Twice* in 1967. 'I have always hated that damned James Bond. I'd like to kill him,' Connery is reported to have exclaimed, perhaps ungratefully, given the fame and fortune it had meant for him personally.

But, as we know, the departure of Connery did not mean the end of 007. Far from it. Producer Cubby Broccoli cast the near unknown George Lazenby to star in *On Her Majesty's Secret Service* in 1969. In retrospect, it was a surprising choice given that his screen exposure had largely comprised an appearance in a black and white Fry's Chocolate commercial. It is however undeniable that Lazenby looked the part and his dark-haired athleticism bore some resemblance to his celebrated predecessor. Unfortunately for 'The Big Fry' (as the press inevitably called Lazenby), his performance drew less than favourable comparisons. He too quit shortly after filming and Sean Connery was persuaded to return one last time for a reputed £1.25 million.

Casting a new James Bond, then, is not as easy as it looks. Whoever is selected is, after all, at the helm or at least the figurehead, of a multi-million-dollar business. Lazenby had his critics, but he at least demonstrated that Bond, who had survived the most improbable and death-defying situations in the movies, could also survive the departure of Sean Connery. We are now used to successive actors passing the baton and the question of 'who will be the next James Bond?' is a source of near endless press speculation.

Roger Moore's tenure throughout the 1970s and into the 1980s might have meant a lighter touch, but it also cemented that survival of the series. Moreover, it showed that the role could adapt and move with the times, appealing to new audiences. In the 'Battle of the Bonds' of 1983, when Connery was lured back in the unofficial film *Never Say Never Again*, it was Moore's *Octopussy* which triumphed at the box office.[48]

It became clear that each Bond could be a little different and periodic recasting could be an opportunity not only for continuity but also for reinvention. That was true of Timothy Dalton, whose two films at the end of the 1980s brought a more brutal realism to the character in place of gadgets and gimmicks. It was also true of Pierce Brosnan, whose suave presence demonstrated that Bond was not only back, after a six-year hiatus, but was also remoulded ready for the twenty-first century. And, of course, it was also true when Daniel Craig was cast for the 2006 reboot, *Casino Royale*. Craig was a controversial and, in some respects, unusual choice. Blonde and also rather craggy, he wasn't an obvious fit. His debut saw a more violent and dark James Bond, in what was a risky departure from the fun action adventures of the Brosnan era.

As it turns out, however, it was a risk that paid off. Craig's Bond movies count among the most commercially successful of the entire franchise. And that intuitive risk-taking is likely to remain in human hands as we face the Fourth Industrial Revolution.

X

On the Road is considered one of the most important novels of the twentieth century. Written by Jack Kerouac, it portrays an underclass of post-war youth culture in an America, where the national narrative was quite different: the narrative of Levittown where Theodore and Patricia Bladykas had settled into the consumerism of the American Dream. *On the Road*, by contrast, was about non-conformity, discontent and living on the edge. It is a novel about the ideas of a young generation faced with Cold War, modernity, class and race struggles. It was influenced by the free flowing bop style of jazz which this Beat Generation associated (Dizzie Gillespie even named a tune 'Kerouac'). That free-flowing prose was a revelation and as controversial as the contents, with its depictions of drug use and sex.

What is extraordinary is that it took Kerouac just three weeks to write *On the Road*. With the story developing in his head over the preceding years, sat at his table drinking coffee and fuelled by Benzedrine, Kerouac typed non-stop on a 120-foot roll of paper, meaning that he did not need to stop the flow to change sheets. The writing might have been quick, but it took a staggering six years for the book to be published, rejected time and time again by fearful and cautious publishers.

Viking eventually published the book, which went on to sell around four million copies in the sixty years since. But more than that, *On the Road* influenced the creative outputs of writers poets and musicians including Patti Smith and Tom Waits, distinctive artists who themselves struggled with publishers and record companies' attempts to contain their originality and creativity in favour of tried and trusted formulas.

To reimagine, to reinvent. For some time at least, that takes human creativity and human judgement. AI alone cannot do this, but it surely has a part to play collaborating with human ingenuity.

XI

'DNA Footprint' is a sophisticated AI system developed by software company Largo. In 2020, the firm put it to good use by asking the question, who will be the next actor to play James Bond?

Focussing on actor-character characteristic correlations, DNA Footprint identified 1,000 actor attributes and cross-compared them with the DNA footprint of James Bond himself, drawn from those six decades of screen appearances. Moreover, it assessed the audience reaction to the potential actors' Bond attributes. Billed as the 'first ever AI assisted casting',[49] this complex system appeared capable of making a scientific decision at a level humans could not hope to achieve.

And the actor 'cast' by DNA Footprint? Well it was Henry Cavill, the tall, dark-haired, handsome thirty-seven-year-old, long a favourite to don the famous tuxedo. It is not hard to picture Cavill as 007, and he was actually screen-tested (aged just twenty-two) for the part in 2005 before Craig was eventually cast. So, does this demonstrate the success of AI or suggest broader lessons for leadership?

Cavill is an easy choice for Bond. His on-screen character attributes correlate with those of the fictional super spy. Audiences are unlikely to be offended by his casting, and producers could be reassured that he could take on the role with confidence. But Cavill would be a safe choice, when the success of the movie franchise over such a long period has been to reflect on the times, update itself and take a step in a slightly different direction by making creative choices. It did that when producer Barbara Broccoli cast Daniel Craig as a very different Bond than his predecessor: different in stature and different in demeanour. One might even say displaying some different attributes.

Would DNA Footprint have cast Craig? One has to doubt it. Just as AI would have likely rejected Kerouac, because that creative jump in viewing something familiar afresh, to build on what has gone before but to take it somewhere new, to produce something distinctive, that is a human attribute that AI alone seems far from being able to replicate. That was part of the message Ai-Da was delivering to Parliament. The technology does not solve the problem that Kerouac faced in struggling to have *On the Road* published. Just like Amazon's recruitment engine, the technology can institution-alise mediocrity; it seeks consistency with the past rather than appreciating the value of creativity and the innovative leap into the new. Like the experience of political social media, it runs the risk of becoming an echo chamber rather than an engine of possibilities. The decision to publish *On the Road*, just as the decision to cast Daniel Craig, needed leadership. But it also needed an understanding of the power of creativity.

XII

The importance of creativity, a uniquely human skill, has run through the stories and arguments in this book. Creativity is an essential component of innovation and has always been a prerequisite for progress, whether that be in business, public service, politics or economics. And as we reach this inflec-tion point, it is likely that the balance between performance management and creativity is tipping firmly in favour of the latter. It is clear that to harness the possibilities of the digital revolution, those very human, unpredictable, inspirational, creative practices need to be allowed to flourish. And that is a challenge for leadership.

The nature of creativity, the use of imagination, the con-nections made between sometimes disparate ideas to produce something new and valuable, is a delicate and mysterious process. The comedian, actor and Professor at Large John Cleese puts it as simply, 'new ways of thinking about things… Creativity can be seen in every area of life – in science, or in business or in sport. Wherever you can find a way of doing things that is better than what has been done before, you are being creative.'[50] It is a distinctly human phenomenon that has been central to human development. Organisationally, it is not something that can easily emerge or reach its true potential in command control environments, and those who are determined to cling on to old industrial practices, mindsets and structures will surely pay the price of disruption. This is what Brian MacDonald must have realised when he and his team decamped to Hawaii to develop Microsoft Teams. He wanted an open, creative, environment where fresh ideas could emerge.

In his delightful book, *Nina Simone's Gum* (itself a work of extraordinary creativity), the instrumentalist and composer Warren Ellis offers some insight into 'the care needed… when an idea is presented, that the wrong words can deflate someone's attempts. That we can easily kill an idea. That fragile moment when it is presented. The permission to proceed given by the confidence of others.'[51] That permission to proceed is like the permission to fail and to exist requires trust and freedom. As Cleese adds: 'Let me reassure you. When you're being creative, there is no such thing as a mistake. The reason is very simple: you can't possibly know if you are going down a wrong avenue until you've gone down it.'[52] That importance of failure to learn and to innovate is of course the very antithesis of approaches like Six Sigma that remain so attractive to too many leaders.

XIII

Creativity flourishes with the freedom from oppression, of authoritarian management and of performance-dominated strategy. But – just like with the Microsoft Teams app, Pomegranate House, the next Bond film, the next Nick Cave and the Bad Seeds album featuring Warren Ellis, or the Electric Telegraph, the 'Taxi de la Marne' story, Participatory Budgeting, Miss Debater, Tec-Tec or Pepper the robot – without leadership it is much less likely to happen. It is here that the concept of creative leadership has emerged and is an approach to leadership which prioritises fostering those very conditions that allow and inspire creative ideas to emerge. It 'can be manifested in the forms of Facilitating, Directing, or Integrating'⁵³ and harnesses shared ambitions and collaboration.

Creativity is not just about art. It stretches into every area of enterprise, science and public life. Creativity is needed to unlock the possibilities of the digital revolution, that will emerge from collaboration between specialists and technology, that will adapt and bridge and that will address the challenges of tomorrow. Human creativity is needed to reclaim the revolution.

That would seem a suitable point to end this chapter and the book.

Conclusion:
Reclaiming the Inflection Point

*'This revolution is happening whether we recognise it or
not and whether we resolve to do anything about it or not'*

When world leaders gathered in Egypt for COP27, the UN
climate change conference, the discussion was decidedly retro-
spective. The acknowledgement, ever present, was summed up
by the UN Secretary General, who told delegates that humanity was on a 'highway to climate hell'. António Guterres
added the warning that 'our planet is fast approaching tipping
points that will make climate chaos irreversible... We are on
a highway to climate hell with our foot on the accelerator.'[1]
The main argument, though, as leaders agreed on ends but
squabbled over means, was the contentious issue of 'loss and
damage'. That is, the countries most affected today by the
ravages of climate change are also those least to blame for
carbon emissions. It is the richer countries who have emitted
the majority of greenhouse gases since industrialisation and,
the argument goes, should pay up.

But one little reported presentation in Egypt that November
was from the team behind DestinE (Destination Earth), the
EU project to create a digital twin of our planet. The ambition
here is for a fully configurable climate information system
able to simulate changes in the environment. It cannot solve

climate change but it can help to understand what is happening, how it might be tackled and the interventions likely to succeed. It will offer evidence, accuracy, speed, responses. COP27, like so much of our public discourse, was stuck in an old industrial mindset, not yet ready to appreciate the multi-dimensional, radical change offered by exponential technology. And the climate emergency, along with transformative emerging technologies like DestinE, appears as something of a metaphor for this inflection point. It represents an inevitability, hurtling towards us, something so significant that needs global collective intervention, where there are avenues of action and revolutionary technologies but where humanity prefers ignorance at worst or to assume the role of passive observers at best.

As extraordinary as the possibilities of the digital revolution before us, what is incredible is that human society appears so unprepared for what is about to happen. There are pockets of impressive progress around the world including Society 5.0[2] in Japan, which acknowledges the need to converge virtual and physical spaces to balance economic development and social issues; the I40 Strategy in Germany, whose mission it is to future-proof manufacturing by integrating technologies,[3] but these are as yet initiatives and not the deep, integrated approach to policy that will be needed. We are at an inflection point of transformation in our society that will change the way we live, interact, work and create. Technology like Pepper the robot or AI-Da, Tec-Tec, Miss Debater, pilotless aircraft, Social Credit, Spot the dog, and many, many more, will transform, interact, learn, and reach into our lives in ways that we have only begun to imagine. This new digital world will change how we work and what humans do in work. It will replace human jobs and change the nature of the professions. Just like the example of Pomegranate House seeking fresh approaches to an industrial age problem, it will free up human capacity to create, to reimagine and to do new things. It will transform our economies and as a result represents a truly significant challenge to politics across the world. This revolution is happening, whether we recognise it or not and whether we resolve to do anything about it or not. It cannot be arrested.

The response of politics to the disruption of the last decade or so, and to the frustration of cynical rational choice electoral strategies, has been the emergence of fantasy politics, populism and disruption. It brought us Trump, Brexit, the AfD, Meloni and many more examples of divisive, anti-establishment movements. Since the global financial crisis the number of populists in power around the world has accelerated sharply to what the Tony Blair Institute reports as a three-decades high.[4] But the disruption of the financial crisis and the period of globalisation preceding it is but a prelude to the coming revolution. Politics needs to wake up.

There are signs that this damaging fantasy politics is at last on the wane – not simply as measured by electoral defeats (which are just as capable of energising a base as a win) but more encouragingly in the collapse of their patent absurdity. A case in point is the UK which, in 2022, had three Prime Ministers and four Chancellors of the Exchequer midway through a Parliament and with a governing party enjoying a sizable majority. When the personal excesses of populist Boris Johnson became too much, and his poll ratings slipped, he was forced from office during a dramatic few days that saw some fifty ministers resign from his government. Not yet done with fantasy, his successor, Liz Truss, and her kindred spirit Chancellor Kwasi Kwarteng, announced a dramatic, populist, tax-cutting 'Mini-Budget', pitching themselves as crusaders in opposition to the establishment 'economic orthodoxy' that they claimed had held the country back. Their unfunded plans led to a collapse in Sterling, a slump in the FTSE and a spike in borrowing on the bond markets. The Bank of England was forced to make an emergency intervention. Kwarteng was sacked and then Truss was forced to resign – at just forty-nine days in office, she became Britain's shortest-serving Prime Minister.

As costly as the episode was, financially and reputationally, it might just have proved to be the point where fantasy politics was shown to be just that – a fantasy. It demonstrated in irrefutable terms that politics and economics have to operate in the real world. It was so disastrous that it was near fatal to right-wing populism in government. A new Chancellor, Jeremy Hunt, reversed Kwarteng's Budget and then announced a hefty rise in taxation to stabilise public finances. A new Prime Minister in Rishi Sunak signalled a return to responsible government and economic discipline. And Jacob Rees-Mogg left Cabinet (a bit like ousted Jeremy Corbyn from the front bench on the other side of the House) returning to the backbenches of the House of Commons, where he resumed his role, from a sedentary position, as critic of any sensible policies in range.

A thread which has run through this book is not simply that the revolution represents a dramatic challenge to democratic politics but also that it represents a challenge to humanity. Industry 4.0 will be driven by technological capability, but it must be harnessed by humanity for human society. We need to shape the revolution in our values. Unfortunately, the disruptive era through which we have been living demonstrates that democratic politics is not yet at the point where it is able to shape the revolution. It is insufficient to defeat fantasy politics if that means simply a return to rational choice politics which themselves failed to understand the disaggregated needs of so many citizens.

The Fourth Industrial Revolution demands inclusive economics combined with a form of deliberative politics that means citizens no longer act as voting consumers but rather take joint responsibility for decisions. Only then will those values be represented and local economies have any hope of exploiting the possibilities of the revolution. Fortunately, the emerging technology offers the possibility even here in harnessing new capabilities, collaborating with people, to strengthen political discourse. Reclaiming the revolution requires the sort of political leadership that embraces deliberation and inclusivity. It needs to shape the society physically and connect it digitally. It needs to invest in education, creativity and lifelong learning. It needs to engage in the really difficult ethical debates that mean machines will behave, learn and interact in accordance with our values. It is clear that politics as usual is unlikely to be capable of achieving this. A new politics is needed at this inflection point capable of understanding and developing the human skills necessary to take advantage of the possibilities and to harness the technology for public good.

What is needed for this to happen? Leadership. Leadership that is brave, visionary, and trusting. Leadership that abandons the discredited political approaches and enables society to shape the inflection point and reclaim the revolution of tomorrow. And while that is true of political leadership, it is also true of leadership throughout public life, industry, public services and other institutions. Technological transformation means that, more than ever, we need to value and nurture the human skills of creativity, adaptability and innovation. The limits of performance management are already there to see. Leadership needs to decouple itself from its obsession with performance and trust dynamic environments, like those found in Building 20, to bring together specialists in unexpected and fruitful ways. After all, Industry 4.0 requires that the worst excesses of command control must be swept aside to allow people to collaborate and make new connections.

And the demand of this inflection point is twofold: it is to tackle the vast new challenges, and more pressing, the need also to revisit the unsolved issues which have plagued societies since industrialisation itself. To take the opportunity to reappraise how we organise, how we share, how we make decisions. It means being motivated to solve problems not to dictate, control and restrict. That, of course, includes the climate crisis, perhaps the biggest challenge of our time, frustratingly discussed by world leaders at COP27.

What have we seen through the adventures contained in these pages? It is that openness, free thinking, inclusiveness, trust and deliberation are the paths to a successful future. And we have seen here how so much of our traditional modes of leadership have got in the way of the creative thinking we have needed up until now. We are at the inflection point of a transformative revolution. It must be reclaimed.

Notes

INTRODUCTION

1   Schwab, K. 2019. Speech to Chicago Council on Global Affairs, 5 June.
2   https://digital-strategy.ec.europa.eu/en/policies/destination-earth.
3   Urbina, F., Lentzos, F., Invernizzi, C. and Ekins, S., 2022. Dual use of artificial-intelligence-powered drug discovery. *Nature Machine Intelligence*, 4(3), pp. 189–191.

PART I

1   Kaye, K. 2022, 'Companies are using AI to monitor your mood during sales calls. Zoom could be next'. *Protocol*, 13 April.
2   Pepper FAQs. https://www.generationrobots.com/pepper/.
3   Mezuk, B., Rock, A., Lohman, M.C. and Choi, M., 2014. 'Suicide risk in long-term care facilities: A systematic review'. *International Journal of Geriatric Psychiatry*, 29(12), pp. 1198–1211.
4   See for instance the tracking of Social Development Scores. https://www.carnegiecouncil.org/studio/multimedia/20130315-the-measure-of-civilization-how-social-development-decides-the-fate-of-nations.
5   For case studies on the process see Billsberry, J. 2008. *Experiencing Recruitment and Selection*. John Wiley & Sons.

6   Bertrand, M. and Mullainathan, S. 2004. 'Are Emily and Greg more employable than Lakisha and Jamal? A field experiment on labor market discrimination'. *American Economic Review,* 94(4), pp. 991–1013.

7   Bigman, Y.E., Wilson, D., Arnestad, M.N., Waytz, A. and Gray, K. 2022. 'Algorithmic discrimination causes less moral outrage than human discrimination', *Journal of Experimental Psychology: General.*

8   World Health Organization. 2021. *Mental Health Atlas* 2020.

9   Chang, B., Gitlin, D. and Patel R. 2011. 'The depressed patient and suicidal patient in the emergency department: evidence-based management and treatment strategies'. *Emergency Medicine Practice*, 13 (9): 1–23, quiz 23–24.

10 Franklin, J.C., Ribeiro, J.D., Fox, K.R., Bentley, K.H., Kleiman, E.M., Huang, X., Musacchio, K.M., Jaroszewski, A.C., Chang, B.P. and Nock, M.K. 2017. 'Risk factors for suicidal thoughts and behaviors: A meta-analysis of 50 years of research'. *Psychological Bulletin*, 143(2), p. 187.

11 Heller, D. 2016. '50 years of research fails to improve suicide prediction: study'. *Medical Xpress*, November 18.

12 Franklin et al., op. cit.

13 BBC World, 2019. 'Predicting Suicide'. https://www.bbc.co.uk/programmes/p071746x

14 Arden, J. 2019, *The Philosopher Lecturing on the Orrery*, Quandary Books.

15 McCloskey, D. 2020. 'The Great Enrichment'. *Discourse*, July 13.

16 Kyodo, 2016, 'Drunken Kanagawa man arrested after kicking SoftBank robot'. *The Japan Times*, Sept 7.

17 World Economic Forum, 2020. 'The Future of Jobs Report 2020'. Geneva, Switzerland: World Economic Forum, October.

18 See https://www.kela.fi/web/en/basic-income-objectives-and-implementation, https://www.theguardian.com/society/2017/feb/19/basic-income-finland-low-wages-fewer-jobs and https://www.hs.fi/talous/art-2000005044824.html?share=9391f3b874e7f9fe70de83a5cb58ebf1.

19 Paine, T. 1796. *Agrarian Justice*.

20 Ibid.

21 See for instance Booth, P. 2018. 'The Case Against a Universal Basic Income', IEA, 22 May.

22 Associated Press, 1992. 'Richman Chain Will Be Closed', April 3.

23 Standing, G. 2012. 'The precariat: from denizens to citizens?' *Polity*, 44(4), 588-608.

24 Verma, P. 2022. 'Humans vs Robots: the battle reaches a 'turning point'', *Washington Post*, December 10.

25 'How Robots Change the World'. *Oxford Economics*, June 2019.

26 Galbraith, J.K., 2014. *The End of Normal: The great crisis and the future of growth*. Simon and Schuster.

27 *Oxford Economics*, op. cit., p. 7.

28 Quartz, 2017. 'The robot that takes your job should pay taxes, says Bill Gates', February 17.

29 'How Experts Think We'll Live in 2000 AD', *Robesonian* (NC) / Associated Press, December 27 1950.

30 Gates, B., 2017. 'Council of the Great City Schools'.

31 Ibid.

32 BLS, 2019. 'Job Openings and Labor Turnover Summary'. US Bureau of Labor Statistics, February 12.

33 SHRM, 2019. 'The Global Skills Shortage'. Society for Human Resource Management, https://www.shrm.org/hr-today/trends-and-forecasting/research-and-surveys/Documents/SHRM%20Skills%20Gap%202019.pdf.

34 Goldin, C. and Katz, L. 2010, *The Race Between Education and Technology*, Belknap Press.

35 UOB Annual Report 2017. https://www.uobgroup.com/AR2017/documents/Full-Annual-Report-2017.pdf.

36 This is not a new debate, see Bell, D. 1973. *The Coming of the Post-Industrial Society*. Basic Books: New York.

37 See https://www.steelcase.com/microsoft-steelcase/#learn-more.

38 Noam Chomsky was based in Building 20 after he joined MIT in 1955.

39 Plenty has been written about Building 20, including: https://www.businessinsider.com/mits-building-20-is-proof-that-only-a-certain-kind-of-brainstorming-really-works-2012-2?r=US&IR=T, http://www.ingeniousgrowth.com/building-20-the-most-innovative-building-the-world-has-ever-known/ and https://www.nytimes.com/1998/03/31/science/last-rites-for-a-plywood-palace-that-was-a-rock-of-science.html.

40 Kieve, J.L. 1973. *The Electric Telegraph: A Social and Economic History*, David and Charles.

41 Beinhocker, E.D. 2006. *The Origin of Wealth: Evolution, complexity, and the radical remaking of economics*. Harvard Business Press.

42 Burbidge, I. and Webster, H. 2019. 'The public sector innovates just as much as the private sector'. RSA, November 16.

43 Bower, J. and Christensen, C. 1995. 'Disruptive Technologies: catching the wave'. *Harvard Business Review*, Jan-Feb.

44 Krishnan, K. 2019. '3 Vital Skills for the Age of Disruption'. *World Economic Forum*, Sept 30.

45 Christensen, C. Raynor, M. and McDonald, R. 2015. 'What is Disruptive Innovation?. *Harvard Business Review*, December.

46 The pace of change is an interesting concept and the academic Danny Dorling has contested that the speed of change in our societies has actually sowed in recent years. Dorling, D., 2020. *Slowdown: The end of the great acceleration – and why it's good for the planet, the economy, and our lives.* Yale University Press.

47 Schumpeter, J. 1942. *Capitalism, Socialism and Democracy.* Routledge.

48 Steinbuch, Y. 2018. 'Jeff Bezos tells employees that Amazon is not too big to fail'. *New York Post*, November 16.

49 Dobler, T. Sniderman, T. Mahto, M. and Ahrens, C. 2020. 'Swim, Not Just Float: driving innovation and new business models through Industry 4.0'. *Deloitte Insights.*

50 Mistreanu, S., 2018. 'Life inside China's Social Credit Laboratory'. *Foreign Policy*, 3, pp. 1-9.

51 Jackson, S. 2022. 'Meta's latest AI chatbot has mixed feelings about CEO Mark Zuckerberg: "It is funny that he has all this money and still wears the same clothes!"'. *Business Insider*, August 5.

52 Walker, C. 2018. 'What Is "Sharp Power"?'. *Journal of Democracy*, 29, no. 3: 9–23.

53 Webber, A. 2020. 'PwC Facial Recognition Tool Criticised for Home Working Privacy Invasion'. *Personnel Today,* June 16.

54 See https://www.europarl.europa.eu/doceo/document/A-8-2017-0005_EN.html?redirect.

55 Weeks, K.P. 2017. 'Every Generation Wants Meaningful Work-but Thinks Other Age Groups Are in it for the Money'. *Harvard Business Review*, 31.

PART 2

1   Purdum, T. 1995. "Tired' Clinton in Tetons For a 17-Day Vacation'. *New York Times*, August 16.

2   Purdum, T. 1995. 'For Clinton and Family, a Vacation of Golf, Hiking and Breathless Views'. *New York Times*, August 21.

3   Klein, J. 2003. *The Natural: The misunderstood presidency of Bill Clinton*. Broadway Books: New York, p. 40.

4   Downs, A. 1957. 'An economic theory of political action in a democracy'. *Journal of Political Economy*, 65(2), pp. 135-150.

5   Altman, L. 1996. 'Clinton, in Detailed Interview, Calls His Health "Very Good"'. *New York Times*, October 14.

6   Morris, D. 2007. *The New Prince: Machiavelli updated for the twenty-first century*. Renaissance Books.

7   Morris, op. cit.

8   Sorman, G., 2009. *Economics Does Not Lie*. Encounter Books.

9   Fukuyama, F. 1989. 'The End of History?'. *The National Interest* (Summer).

10 Sakwa, R. 2005. *The Rise and Fall of the Soviet Union*. Routledge.

11 Kay, J. 2004, *The Truth About Markets*. Penguin, p. 308.

12 Williamson, J. 1990, 'What Washington Means by Policy Reform', in J. Williamson, ed. *Latin American Adjustment: How Much Has Happened?*. Institute for International Economics.

13 Marshall, C. 2015. 'Levittown, the prototypical American suburb – a history of cities in 50 buildings, day 25'. *The Guardian*, 28 April.

14 *Time Magazine*. 1950. 'Up from the Potato Fields', July 3.

15 *Time Magazine.* op cit.

16 Montgomerie, J. 2009. 'The pursuit of (past) happiness? Middle-class indebtedness and American financialisation'. *New Political Economy*, 14(1), pp. 1-24.

17 Sherwell, P. 2008. 'Financial crisis hits town built for the American Dream', October 11.

18 Stiglitz, J. 2002. *Globalization and its Discontents*, Penguin.

19 Summary of the United Nations Monetary and Financial Conference, July 7, 1944.

20 Zizek, S. 2009. *First as Tragedy, Then as Farce*. Verso, p. 70.

21 Hirschman, A.O., 1970. *Exit, Voice, and Loyalty: Responses to decline in firms, organizations, and states* (Vol. 25). Harvard University Press.

22 Resnick, B. 2017. 'Trump supporters know Trump lies. They just don't care.' *Vox*, July 10

23 Howarth, A. 2018, 'Flashback: When Jacob Rees-Mogg campaigned in Fife with his nanny'. *The Scotsman*, 31 August.

24 Long, C. 2010. 'Maybe he's canvassing in the King of Spain's private loo'. *The Times*, April 10.

25 Müller, J.W.. 2016. *What is populism?* University of Pennsylvania Press; Eatwell, R. and Goodwin, M., 2018. *National Populism: The revolt against liberal democracy*. Penguin UK.

26 McSmith, A. 1999. 'How a point of principle tore our lives apart'. *Observer*, May 16.

27 Wainwright, H. 2018. 'The Remarkable Rise of Jeremy Corbyn'. *New Labor Forum* (Vol. 27, No. 3, pp. 34-42). Sage CA: Los Angeles.

28 Mance, H. and and Pickard, J. 2016, 'Tony Blair refused to expel rebel heir Jeremy Corbyn'. *Financial Times*, September 16.

29 Watts, J. and Bale, T. 2019. 'Populism as an intra-party phenomenon: The British Labour party under Jeremy Corbyn'. *The British Journal of Politics and International Relations*, 21(1), pp. 99-115.

30 Lee, P. 2013. *Nuneaton & Bedworth: Coal, Stone, Clay and Iron*. Amberley Publishing.

31 Dorey, P., 2021. '"David Cameron's catastrophic miscalculation: The EU Referendum, Brexit and the UK's 'culture war"'. *Observatoire de la société britannique*, (27), pp. 195-226.

32 Barber, S. 2017. 'The Brexit environment demands that deliberative democracy meets inclusive growth'. *Local Economy*, 32(3), pp. 219-239.

33 Electoral Reform Society. 2019. 'Revealed: Nearly 200 seats haven't changed hands since World War II', December 2.

34 Parris, M. 2017. 'Awful Rees-Mogg is anything but a joke', *The Times*, August 12.

35 Mance, H. & Pickford, J. 2016, 'Tony Blair refused to expel rebel heir Jeremy Corbyn'. *Financial Times*, September 16.

36 Abbot, D. 2021. 'Andy Burnham: unlikely heir to the left's leadership ambitions?' *The Guardian*, May 19.

37 Hofstede, G. 2009. Geert Hofstede cultural dimensions.

38 Picciotto, R. 2019. 'Is evaluation obsolete in a post-truth world?' *Evaluation and Program Planning*, 73, pp.88-96.

39 Bower, T. 2020. *Boris Johnson: The Gambler*. Random House.

40 Ford, R., Bale, T., Jennings, W. and Surridge, P. 2021. *The British General Election of 2019*. Springer Nature.

41 Siddique, H. 2019, 'Tony Blair says Tories and Labour engaged in "populism running riot"'. *The Guardian*, November 25.

42 Klasen, S., 2010. 'Measuring and monitoring inclusive growth: Multiple definitions, open questions, and some constructive proposals'. McKinley, T., 2010. 'Inclusive growth criteria and indicators: An inclusive growth index for diagnosis of country progress'.

43 Royal Society of Arts, 2016. 'Inclusive Growth Commission: Emerging Findings'. London: RSA. p 3.

44 Stiglitz, J.E., 2015. 'Inequality and Economic Growth'. *The Political Quarterly, 86,* pp. 134–155.

45 Klasen, op. cit.

46 RSA, op. cit. p. 7.

47 Brownstein, R. 2019. 'Biden's been in politics longer than any US presidential nominee ever. Here's why that matters'. *CNN Politics*, May 14.

48 Barber, S. (2014). 'Arise, Careerless Politician: The rise of the professional party leader'. *Politics*, 34(1), 23-31.

49 Ramanujam, S. 2019. 'Harish Natarajan: a human debater who beat a robot'. *The Hindu*, February 20.

50 Campbell, M., Hoane Jr, A.J. and Hsu, F.H., 2002. 'Deep blue'. *Artificial Intelligence*, 134(1-2), pp. 57-83.

51 May, A.M., 1990. 'President Eisenhower, economic policy, and the 1960 presidential election'. *The Journal of Economic History*, 50(2), pp. 417-427.

52 Mattes, K., Spezio, M., Kim, H., Todorov, A., Adolphs, R. and Alvarez, R.M., 2010. Predicting election outcomes from positive and negative trait assessments of candidate images. *Political Psychology, 31*(1), pp. 41-58.

53 Sorenson, T. 2010. 'When Kennedy Met Nixon: the real story', *New York Times*, 25 September

54 Manifesto, L.P. 1997. *New Labour because Britain deserves better.* Labour Party, London.

55 Hall, P.A. 1993. 'Policy paradigms, social learning, and the state: the case of economic policymaking in Britain'. *Comparative politics*, pp. 275-296.

56 World Economic Forum Annual Meeting 2010.

57 Patomäki, H. and Teivainen, T. 2004. 'The World Social Forum: an open space or a movement of movements?'. *Theory, Culture & Society*, 21(6), pp. 145-154.

58 Haidt, J. 2012. *The Righteous Mind: Why good people are divided by politics and religion*. Vintage.

59 Wardle, C. and Derakhshan, H. 2017. 'Information disorder: Toward an interdisciplinary framework for research and policymaking'. Council of Europe.

60 Schwab, Klaus, 2011. 'Shared Norms for a New Reality',. World Economic Forum Annual Meeting.

61 Zuckerberg, M., Dorsey, J. and Pichai, S., 2021. US House Hearing on 'Disinformation Nation: Social Media's Role in Promoting Extremism and Misinformation'.

62 Sunstein, C.R. 2018. 'Is social media good or bad for democracy?' *SUR- International Journal on Human Rights*, 27, p. 83.

63 Barberá, P., Jost, J.T., Nagler, J., Tucker, J.A. and Bonneau, R. 2015. 'Tweeting from left to right: Is online political communication more than an echo chamber?' *Psychological science*, 26(10), pp. 1531-1542.

64 Frimer, J.A., Skitka, L.J. and Motyl, M. 2017.' Liberals and conservatives are similarly motivated to avoid exposure to one another's opinions'. *Journal of Experimental Social Psychology*, 72, pp. 1-12.

65 Colley, D.F. 2018. 'Of Twit-Storms and Demagogues: Trump, Illusory Truths of Patriotism, and the Language of the Twittersphere'. *President Donald Trump and His Political Discourse*. Routledge, pp. 33-51.

66 Dennis, J. 2019. *Beyond Slacktivism*. Cham: Springer International Publishing.

67 https://www.blackrock.com/corporate/investor-relations/larry-fink-chairmans-letter.

68 Chambers, S. 2003. 'Deliberative democratic theory'. *Annual Review of Political Science*, 6(1), pp. 307-326.

69 Habermas, J. 2015. *Between facts and norms: Contributions to a discourse theory of law and democracy*. John Wiley & Sons.

70 Dryzek, J.S. and Braithwaite, V. 2000. 'On the prospects for democratic deliberation: Values analysis applied to australianpolitics'. *Political Psychology*, 21(2), p. 242.

71 Mansbridge, J., Bohman, J., Chambers, S., Christiano, T., Fung, A., Parkinson, J., Thompson, D.F. and Warren, M.E. 2012. 'A systemic approach to deliberative democracy'. *Deliberative systems: Deliberative democracy at the large scale*, pp. 1-26. Owen, D. and Smith, G. 2015. 'Survey article: Deliberation, democracy, and the systemic turn'. *Journal of Political Philosophy*, 23(2), pp. 213-234. Parkinson, J. 2006. *Deliberating in the real world: Problems of legitimacy in deliberative democracy*. Oxford University Press on Demand.

72 LeDuc, L. 2015. 'Referendums and deliberative democracy'. *Electoral Studies*, 38, p. 139.

73 Stevenson, H. and Dryzek, J.S. 2012. 'The legitimacy of multilateral climate governance: A deliberative democratic approach'. *Critical Policy Studies*, 6(1), p. 27.

74 Souza, C. 2001. 'Participatory budgeting in Brazilian cities: limits and possibilities in building democratic institutions'. *Environment and Urbanization*, 13, no.1, pp. 159-184.

PART 3

1   Sharman, J. 2017. 'Grenfell Tower fire: "We woke them up to die" says witness who tried to warn residents of danger', *The Independent*, June 19.

2   Grenfell Tower Regeneration Project Planning Application, October 2012, p. 14.

3   Ibid., p. 15.

4   Graham-Harrison, E. 2017. 'Grenfell fire: The community is close knit – they need to stay here to recover', *The Guardian*, July 9.

5   Bennis, W.G. 2009. *On becoming a leader*. Basic Books.

6   Maxwell, J.C. 2019. *Leadershift: The 11 essential changes every leader must embrace*. HarperCollins Leadership.

7   Proctor, K. 2019. 'Rees-Mogg Sorry for Saying Grenfell Victims Lacked Common Sense'. *The Guardian*, November 5.

8   Jeffery, D., Heppell, T., Hayton, R., & Crines, A. 2018. 'The Conservative Party leadership election of 2016: an analysis of the voting motivations of Conservative parliamentarians'. *Parliamentary Affairs*, 71(2), 263-282.

9   Mathieu, C., Neumann, C.S., Hare, R.D. and Babiak, P. 2014. 'A dark side of leadership: Corporate psychopathy and its influence on employee well-being and job satisfaction'. *Personality and Individual Differences*, 59, pp. 83-88.

10 Anderson, E. and Jamison, B. 2015. 'Do the top US corporations often use the same words in their vision, mission and value statements?', *Journal of Marketing and Management*, 6(1), p.1.

11 Taleb, N.N. 2007. *The Black Swan: The impact of the highly improbable*, Vol. 2, Random House.

12 Hymes, C. 2021. 'Boris Johnson promises 'fluorescent-jacketed chain gangs' so criminals can visibly pay debt', *The Telegraph*, July.

13 Bassett, K. 2017. 'Timpson's chairman on "upside-down management". *Management Today*, June 1.

14 Butler, S. 2019. 'The Support Never Stops – former prisoner working for Timpsons'. *The Guardian*, April 6.

15 Edelman Trust Barometer. 2020. https://www.airmic.com/node/41281.

16 Zak, P.J. 2017. 'The neuroscience of trust'. *Harvard Business Review*, 95(1), p. 84-90.

17 Aksoy, C.G., Eichengreen, B. and Saka, O. 2020. 'The political scar of epidemics'. Available at SSRN 3625190.

18 Zak. 2017. op cit.

19 Editorial. 2017. 'The Guardian view on Grenfell Tower: Theresa May's Hurricane Katrina'. *The Guardian*, June 15.

20 May, T. 2018. 'I made mistakes but one year on I'm going Green for Grenfell'. *Evening Standard*, June 11.

21 Gay, V. 2010. 'Bush/Lauer Part II'. *Newsday*, November 5.

22 Trimmins, N. 2007, 'Public service targets are here to stay'. *Financial Times*, June 25.

23 Wilson, G. 2006. 'Brown has Imposed a Target a Day Since 1997'. *The Telegraph*, December 12.

24 Francis, R. 2013. *Report of the Mid Staffordshire NHS Foundation Trust public inquiry: executive summary*. Vol. 947). The Stationery Office.

25 Goodhart, C. 1975. 'Problems of Monetary Management: The U.K. Experience'. In Courakis, Anthony, S. (ed.). *Inflation, Depression, and Economic Policy in the West. Totowa, New Jersey: Barnes and Noble Books* (published 1981); Strathern, M. 1997. '"Improving ratings": audit

in the British University system'. *European Review*. John Wiley & Sons. 5 (3), pp. 305-321.

26   Hanc, J. 2014. 'A Fleet of Taxis Did Not Really Save Paris From the Germans During World War I: The Myth of the Battle of the Marne has Persisted, but What Exactly Happened in the First Major Conflict of the War?' *Smithsonian. com*, 24.

27   Kwak, Y.H. and Anbari, F.T. 2006. 'Benefits, obstacles, and future of six sigma approach'. *Technovation*, 26(5-6), pp. 708-715.

28   Stulp, G., Buunk, A.P., Verhulst, S. and Pollet, T.V. 2013. 'Tall claims? Sense and nonsense about the importance of height of US presidents'. *The Leadership Quarterly*, 24(1), pp. 159-171.

29   Hanc, J. op cit.

30   Hougaard, R. and Carter, J. 2018. 'Ego is the enemy of good leadership'. *Harvard Business Review*.

31   Sinek, S. 2014. *Leaders eat last: Why some teams pull together and others don't*. Penguin.

32   Hougaard, R. 2018. 'The real crisis in leadership'. *Forbes*.

33   Welch, J. and Byrne, J.A. 2003. 'Jack: Straight from the gut'. *Business Plus*.

34   Kantor, J. and Streitfeld, D. 2015. 'Inside Amazon: Wrestling Big Ideas in a Bruising Workplace'. *The New York Times*, August 15.

35   McLeod, S.A., 2007. 'The Milgram Experiment'. *Simply Psychology*.

36   Giumetti, G.W., Schroeder, A.N. and Switzer III, F.S. 2015. 'Forced distribution rating systems: When does "rank and yank" lead to adverse impact?' *Journal of Applied Psychology*, 100(1), 180.

37 Revzin, S. and Revzin, V. 2018. 'How to Encourgage Entrepreneurial thinking on Your Team'. *Harvard Business Review*, December 21.

38 Warren, T. 2017. 'How Microsoft Built its Slack Competitor'. *The Verge*, March 14. https://www.theverge.com/2017/3/14/14920892/microsoft-teams-interview-behind-the-scenes-slack-competition.

39 https://insideevs.com/news/598270/pedestrians-attack-waymo-self-driving-cars-arizona-california/.

40 Awad, E., Dsouza, S., Kim, R., Schulz, J., Henrich, J., Shariff, A., Bonnefon, J.F. and Rahwan, I. 2018. 'The Moral Machine Experiment'. *Nature*, 563(7729), pp.59-64.

41 McCloskey, D. 2020. 'The Great Enrichment'. *Discourse*, July 13.

42 Blair, T. 2021. 'Without Total Change Labour Will Die', *New Statesman*, May 11.

43 Van Dick, R., Lemoine, J.E., Steffens, N.K., Kerschreiter, R., Akfirat, S.A., Avanzi, L., Dumont, K., Epitropaki, O., Fransen, K., Giessner, S. and González, R. 2018. 'Identity leadership going global: Validation of the Identity Leadership Inventory across 20 countries'. *Journal of Occupational and Organizational Psychology*, 91(4), pp. 697-728.

44 Shaw, D. 2021. 'CEO Secrets: My Billion Pound Company has no HR Department'. BBC, February 24.

45 Hunt, V., Layton, D. and Prince, S. 2015. 'Diversity matters'. *McKinsey & Company*, 1(1), pp. 15-29.

46 Satell, G. and Windschitl, C. 2021. 'High-performing teams start with a culture of shared values'. *Harvard Business Review*.

47 https://www.pomegranatehouse.org/.

48   Ciucchi, B. 1997. 'The Battle of the Bonds'. *Rev Mens Suisse Odontostomatol.*

49   http://largofilms.ch/henry-cavill-swaps-his-cape-for-a-martini-in-ai-victory-as-the-next-bond/.

50   Cleese, J. 2020. *Creativity: a short and cheerful guide.* Crown, p. 3.

51   Ellis, W. 2021. *Nina Simone's Gum*, Faber & Faber, p. 152.

52   Cleese op. cit. p. 48.

53   Mainemelis, C., Kark, R. and Epitropaki, O. 2015. 'Creative leadership: A multi-context conceptualization'. *Academy of Management Annals*, 9(1), pp. 393-482.

CONCLUSION

1   Stanczyk, A. 2022. 'Getting to COP27: Bridging Generational Divide'. *Development*, 65(1), pp. 42-47.

2   https://www8.cao.go.jp/cstp/english/society5_0/index.html.

3   https://ati.ec.europa.eu/reports/policy-briefs/. germany-industry-40.

4   Meyer, B. 2021. 'Populists in Power: Perils and Prospects in 2021'. Tony Blair Institute for Global Change, Oct. 19.

Index

Ingram Content Group UK Ltd.
Milton Keynes UK
UKHW020634180523
421946UK00010B/265